Notified in A.C.Is. 20th May, 1944

226 Publications 52

RESTRICTED

The information given in this document is not to be communicated, either directly or indirectly, to the Press or to any person not authorized to receive it.

HANDBOOK OF ENEMY AMMUNITION

PAMPHLET No. 11

GERMAN MINES, GRENADES, GUN AMMUNITION AND MORTAR AMMUNITION.

By Command of the Army Council,

THE WAR OFFICE,
    20th May, 1944.

HANDBOOK
OF
ENEMY AMMUNITION

CONTENTS TABLE

German Ammunition

	Page
Tellermine 43 (Mushroom) ..	1
Light Anti-tank mine (l.Pz.) ..	4
Hollow charge rifle grenade ..	7
Large hollow charge rifle grenade ..	9
Fuze l.Igr.Z. 23nA ..	13
Fuze A.Z. 35 K. ..	13
Fuze, mechanical, time and percussion, S/90/45 (Dopp. Z.S./90/45) ..	16
Fuze A.Z.2492 ..	23
Fuze A.Z.5072 ..	25
Fuze A.Z.5075 ..	26
Fuze Bd.Z.5130 ..	28
Gaine 34 (Kl.Zdlg. 34 Np) ..	30
Gaine 34 with delay (Kl.Zdlg. 34 Np.mV.) ..	31
Smoke Box No. 8 for H.E. shell ..	31
Smoke Box No. 9 for H.E. shell ..	33
Tracer No. 3 ..	33
2·7 cm. pistol H.E. cartridge ..	35
2·7 cm. pistol signal cartridge ..	37
2·8 cm. Pz.B.41 cartridge Q.F., H.E. ..	39
3·7 cm. Pak cartridge Q.F., H.E./T. ..	41
3·7 cm. Pak hollow charge, muzzle stick bomb (Stielgranate 41)	44
8 cm. M.L. mortar H.E. bomb 38 (Wgr. 38) ..	47
8 cm. M.L. mortar H.E. bomb 39 (Wgr. 39) ..	50
10 cm. M.L. mortar H.E. bomb (Wgr. 35) ..	52
10·5 cm. high velocity H.E. shell ..	54
15 cm. high velocity H.E. shell ..	56

HANDBOOK
OF
ENEMY AMMUNITION

GERMAN TELLERMINE 43 (MUSHROOM) WITH IGNITER T.Mi.Z.42
(T.Mi.Pilz 43/T.Mi.Z.42)

(Fig. 1)

This model of the Tellermine, known as the mushroom type presumably because of its cover plate, differs from the 42 model, described in pamphlet No. 9, mainly in the design of the cover plate. The mine, like the earlier models, is functioned by pressure applied to a contact igniter or by detonators in the side and base, which may be initiated by "booby trap" devices such as the pull igniter Z.Z.35 or by instantaneous detonating fuze connected to another mine. The igniter used, the T.Mi.Z.42, is the same as that described for the Tellermine 42 in Pamphlet No. 9 and is inserted below the cover plate.

The mine is 12·3 inches in diameter and has a convex head to which is fitted a three-tier circular cover plate with a maximum diameter of 7·5 inches at the base. The overall height of the mine with the cover plate is 3·6 inches. The weight of the mine with its amatol bursting charge and P.E.T.N./wax exploders is approximately 17 lb. 5 oz. The exterior is painted grey and is stencilled on the top, in white, with the abbreviated designations of the mine and igniter, "T.Mi.Pilz 43/T.Mi.Z.42. 13A". The marking "13A", following the marking relating to the fuze, indicates the bursting charge to be amatol 50/50.

Body

The steel body consists of two main parts, the charge container and the base plate. The base plate closes the base of the container and is provided with a stepped rim which fits inside the container and is secured by the lower edge of the container being turned in to engage the step. A socket for a detonator and pull igniter is fitted in the base plate. A screwthread is formed for the insertion of the pull igniter. The socket is secured by two clips which engage in a slotted flange at its base.

The steel charge container is fitted with a side lifting handle and may have a strip shaped to form a loop fitted also on the side. The top of the container slopes upwards towards the centre, where it is flat with a groove encircling the fuze hole. Inside the container,

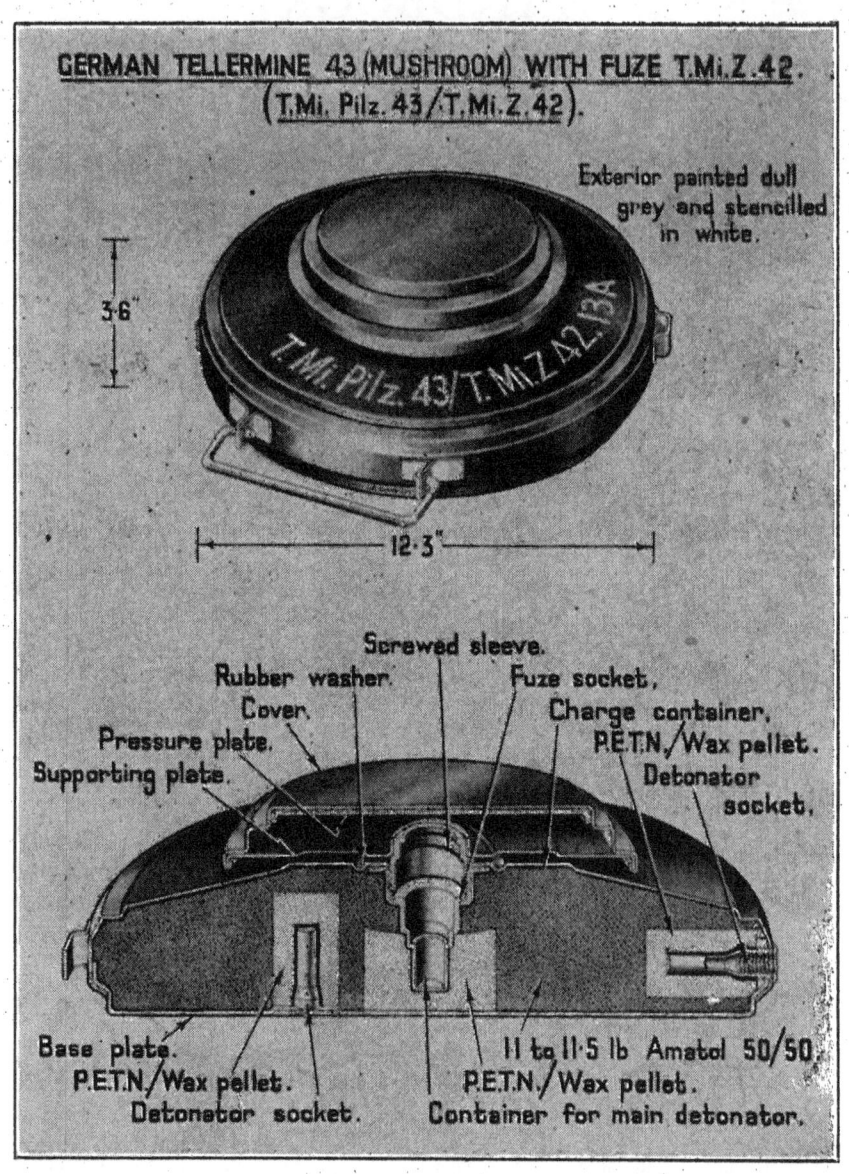

Fig. 1

below the fuze hole, a steel socket is fitted to receive the igniter. The socket is screwthreaded internally for the assembly of the cover plate and is fitted with a container at its base to accommodate the main detonator. A socket for the insertion of a detonator with a pull igniter is fitted in the side of the container.

Cover Plate Assembly

The cover plate is attached to the body by a screwthreaded sleeve, protruding from the centre of its base, which is screwed into the igniter socket in the top of the body. A rubber washer seals the joint between the top of the body and the underside of the cover plate. The cover plate assembly consists of a light pressed steel circular cover, shaped to form three tiers, with a pressure plate secured inside the top tier and a supporting plate closing the base. The screwed sleeve for the assembly of the cover plate with the body is fitted in the supporting plate.

The diameters of the three tiers of the cover are approximately 7·5 inch at the base, 6·7 inch at the centre, and 5·9 inch at the top, the heights being approximately ·3 inch, ·5 inch and ·4 inch respectively. The pressure plate is held inside the top tier by spot welding at points on its circumference and consists of a mild steel circular plate with a circumferential rim on its underside. The supporting plate is held inside the bottom tier by spot welding, in the same way as the pressure plate, and by the rim at the base of the cover being turned in under a rim on the base of the plate. The plate is of mild steel with a circular corrugation and with a rimmed hole in the centre in which the screwthreaded sleeve is secured by spot welding.

Bursting Charge and Exploders

The interior of the body is painted black before filling. The bursting charge consists of 11 to 11·5 lb. of Amatol 50/50 filled from the base to a density of 1·4. An unwrapped exploder pellet of P.E.T.N./wax 89/11 is inserted to form a surround for each of the detonator holders during the process of filling. The weights and densities of the exploder pellets are :—

    Exploder for main detonator 4¼ oz. density 1·57

    Exploder for side detonator 1¾ oz. density 1·56

    Exploder for base detonator 1¾ oz. density 1·56

Main Detonator

The main detonator is attached to the igniter and is described, with the igniter, in Pamphlet No. 9.

Action

The cover plate is removed to insert the igniter with the main detonator. When a load is applied to the cover plate the thin wall of the cover collapses and the pressure plate is forced down on the striker of the igniter, with the result that the shear pin is severed and the fuze functions. The load required to function the mine with the T.Mi.Z.42 igniter will be approximately the same as for the Tellermine 42, that is, 570 lb.

The mine can also be detonated by pull igniters, which function if an attempt is made to lift the mine before these are disconnected.

The pull igniter Z.Z.35 is described in Pamphlet No. 8. The position of the sockets for these igniters in the side and base of the mine have been found to vary. Regarding the centre of the lifting handle as twelve o'clock, the side socket may be found between the positions of two and three o'clock or at six o'clock. The base socket is about 4 inches in from the periphery and may be found at positions corresponding to one o'clock or between eight and nine o'clock. There may, of course, be further variations.

GERMAN LIGHT ANTI-TANK MINE
(1.Pz.)
(Fig. 2)

The mine is designed specially for use by airborne troops and is functioned by pressure applied to push igniters, of which there are five fixed between the top and base of the outer casing. By removing the nuts holding the igniters at the underside of the base and placing the mine so that the protruding lower ends of the igniters are supported on an unyielding surface, the mine can be used as an anti-personnel mine. Sockets for insertion of anti-lifting devices, such as the Z.Z.35 used with the Tellermines, are not provided.

For transport purposes the mines are packed in crates which hold five. The crates are packed into containers for dropping.

The flat circular outer casing of the mine with its side radiussed at the top and bottom is approximately 10·3 inches in diameter and 2·4 inches in height. A safety plug with a milled head protrudes from the centre of the top of the casing. At equally spaced points around the top of the casing there are three small hexagonal nuts and, nearer the side, five larger hexagonal screwed plugs. The milled head of the safety plunger is engraved with the word "Sicher" (Safe), an arrow indicating the direction in which the head should be turned to make the mine safe and an index line which must be coincident with a mark on the top of the casing when the safety plug has been screwed fully home into the safe position. The exterior of the mine is painted grey on deep olive green and the overall height is 3·5 inches. The weight of the complete mine with its 4·75 lb. bursting charge of T.N.T. is 9 lb. The use of a small metal cover to fit over the safety plug has been reported.

Body

The mine consists of a pressed steel outer casing which encloses the charge container and is in two parts, the top part being lipped to fit tightly over the rim at the top of the base part. The joint between the two parts is made waterproof with adhesive tape after the mine has been assembled. Attached to the underside of the centre of the top part, by means of five screws, is the flash chamber

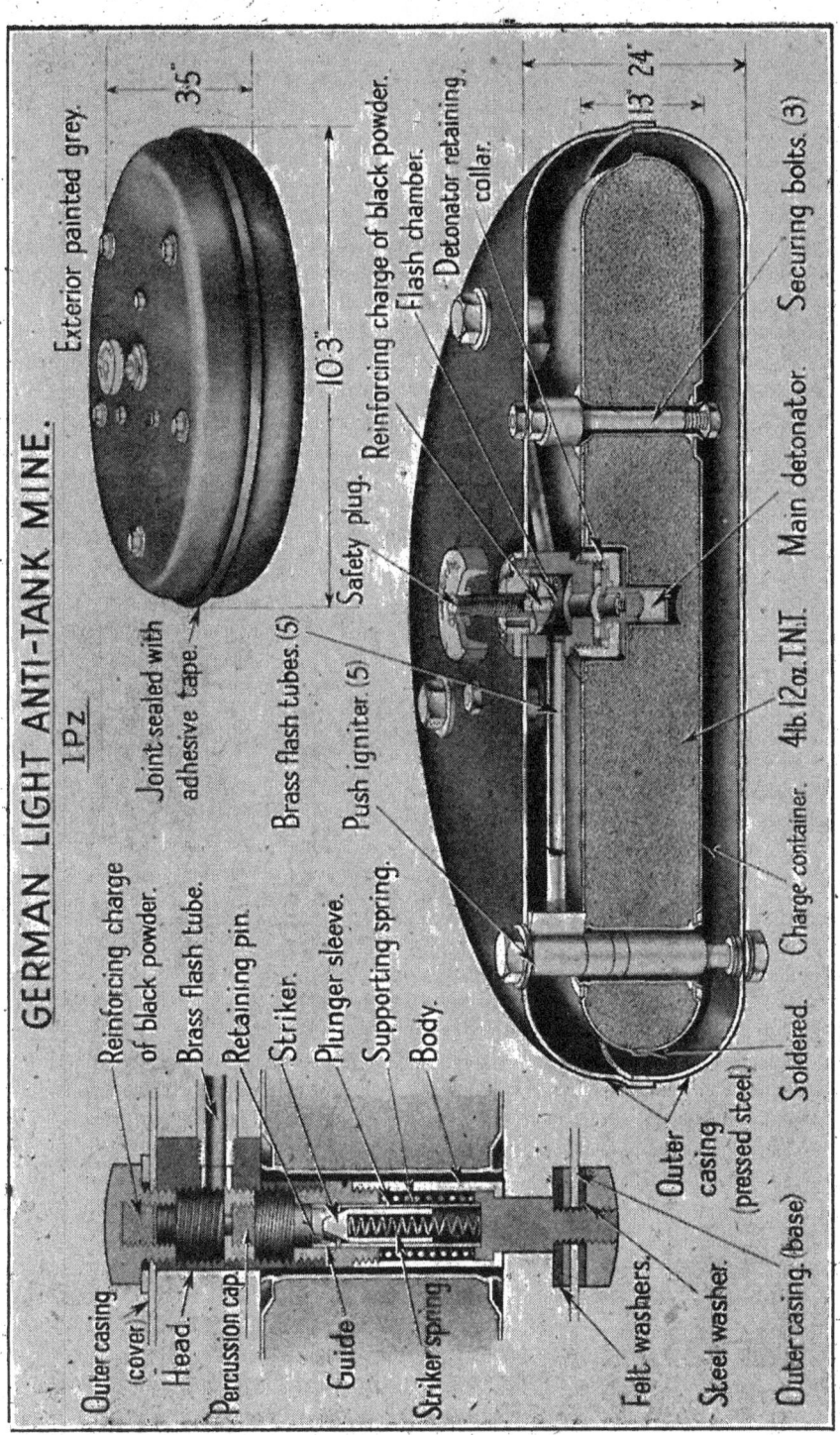

FIG. 2

with the detonator retaining collar screwed to its base. Five brass radial flash tubes connect the flash chamber with the heads of five push igniters. A hole in the base of the flash chamber leads to the main detonator attached beneath. A circular plate on the top of the casing above the flash chamber has a screwthreaded hole in the centre in which the screwed safety plug is inserted. The lower end of the safety plug is enlarged and coned so that when it is screwed in it closes the hole in the base of the flash chamber. The coned head is recessed at the centre of its underside, the recess being filled with a reinforcing charge of black powder.

The charge container of tinplate is similar in shape to the outer casing inside of which it is supported, with air spacing all around it, by three bolts. The bolts pass through tubes fixed between the upper and lower parts of the container and are secured by nuts to the top of the outer casing and the base of the container. Five similar but larger tubes are fitted in the container for the push igniters. A socket for the main detonator is fitted at the centre of the top part of the container. The container is completely sealed with solder of low melting point.

Bursting Charge and Main Detonator

The bursting charge consists of 4 lb. 12 oz. of a high grade T.N.T. having a setting point of 80·6 degrees Centigrade. The density of the filling varies from 1·42 at the outer edge to 1·48 around the detonator recess. The T.N.T. is probably milled and pressed in increments.

The main detonator is the same as that used in the Tellermine 35 and is described in Pamphlet No. 9.

Push Igniter

The five igniters operate on the same principle as the S.Mi.Z.35 used in the S-mine. The spring-loaded striker is held in position by two small pins housed in holes in the sleeve of the plunger where they are retained by the guide. The plunger is nutted to the base of the outer casing, whilst the body, with its head which contains a percussion cap, is secured to the top of the outer casing by a screwed plug. The screwed plug contains a reinforcing charge of black powder pressed into a recess in its base. With the nut removed from the plunger at the base of the outer casing, the plunger is held away from the guide only by the supporting spring.

Action

The safety plug must be unscrewed to its full extent to unmask the hole in the base of the flash chamber when required for use. When the hexagon nuts beneath the base part of the outer casing are screwed up tightly the mine is in condition for use as an anti-tank mine and functions when the casing is crushed sufficiently to operate one of the push igniters. When the top of the casing is forced downwards it takes with it the igniter, except the plunger, which is supported by the base part of the casing and contains the spring-loaded striker in its sleeve. When the guide in the igniter has moved downwards sufficiently to unmask the holes in the sleeve containing the retaining pins, the pins are forced outwards by the inclined surface on the striker and the pressure of the striker spring, which then drives the striker into the percussion detonator. The flash from the detonator, reinforced by the charge in the screwed plug above, passes through the brass flash tube to the flash chamber. In the chamber the flash is again reinforced by the charge in the coned head of the safety plug and enters the main detonator through the hole in the base of the chamber. The weight required to function the mine in this manner is not yet known.

When required for use as an anti-personnel mine, the hexagonal nuts securing the igniter plungers beneath the base of the casing are removed and the mine is placed in position so that the protruding base ends of the plungers are supported on an unyielding surface. When pressure is applied, the outer casing carrying the igniters is forced downwards compressing the supporting spring in the igniter, whilst the plunger is supported by the surface on which its base end is resting. The protrusion of the base end of the plunger is such that it permits the casing, with the attached igniter, to move downwards sufficiently over the plunger to result in the unmasking of the holes in the sleeve which contain the retaining pins and the striker is released. The actual load required will depend upon the point of application and the surface on which the mine is resting. It may even be as low as that needed to compress one igniter supporting spring. Tests carried out on two igniters taken from a mine showed that the functioning load for a single igniter is from 10 to 12 lb.

GERMAN HOLLOW CHARGE RIFLE GRENADE
(Gewehr Panzergranate)
(Fig. 3)

The grenade is fired from the 3 cm. rifled discharger cup which can be fitted to the 7·92 mm. Mauser rifle. The propellant cartridge is crimped at the mouth.

The rustproofed steel body of the grenade is cylindrical in shape, with a parabolic steel impact cap at the head and a tail unit, of the same diameter, screwed on at the base. The tail unit is of unpainted aluminium and has a pre-engraved band near the base with eight grooves. The overall length of the grenade is 6·4 inches and the diameter of the body, 1·175 inches. The total weight is 8 oz. 5 drs.

The body contains a 1 oz. 12 dr. bursting charge of T.N.T. with a cone shaped " hollow " in which there is a steel cavity liner.

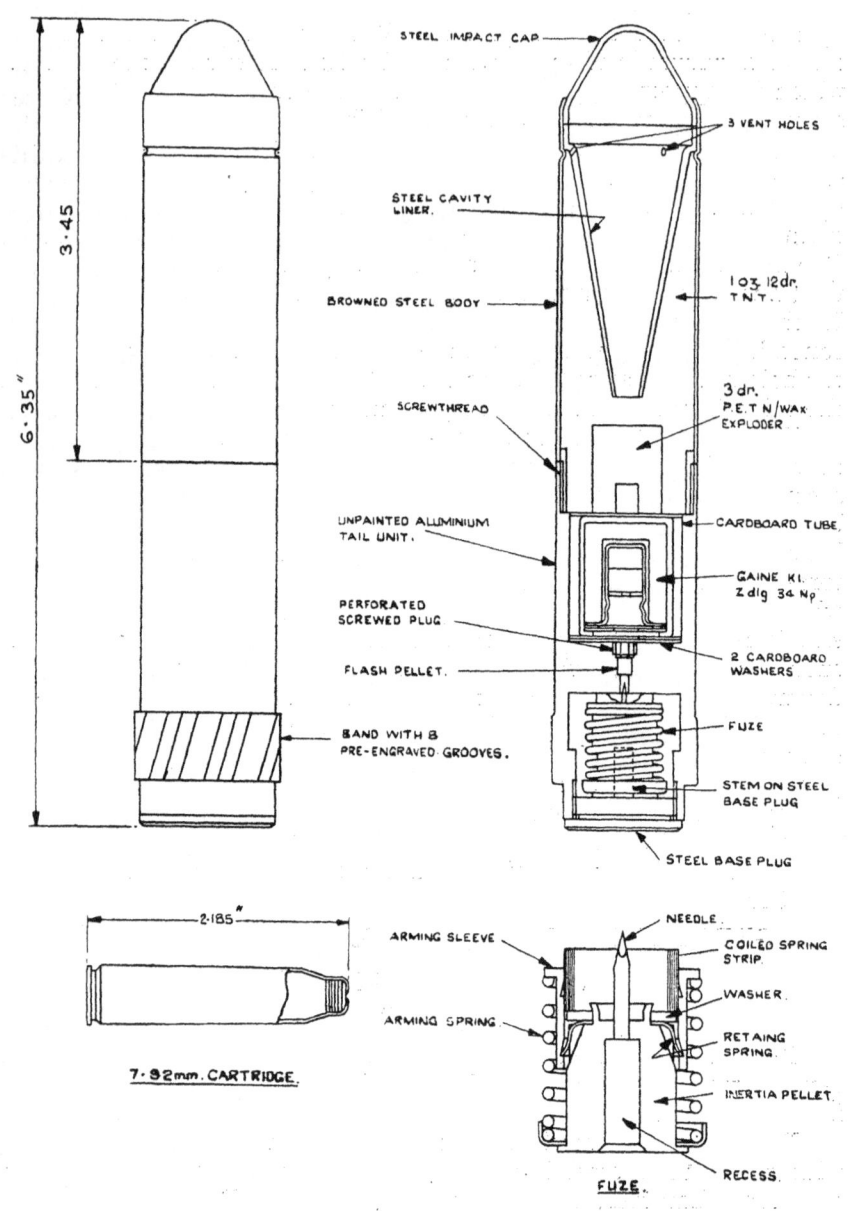

Fig. 3

The liner is coned at 19·5 degrees and has the apex cut away. The top of the liner is cylindrical for assembly in the body and has three equally spaced holes, presumably for the escape of air during filling. The thickness of the wall of the liner tapers from ·047 inch at the top to ·015 inch at the base. A 3 dram P.E.T.N./wax exploder is contained in a cavity in the base of the bursting charge. The steel impact cap is secured above the cavity liner by the spun body. At the base end of the body a screwthreaded tube is fitted for the assembly of the tail unit.

The tail unit with its gaine and fuze is the same as that used with the large grenade.

Cartridge

The cartridge consists of a 7·92 mm. S.A. case of steel, coated with gilding metal or varnished, containing a 13 grain propellant charge of nitrocellulose powder. The mouth of the case is crimped and waxed over a fibre wad. The propellant is the same as that described for the large hollow charge rifle grenade.

Perforation of Armour

Perforation of 38·5 mm. homogeneous hard armour has been obtained at 30 degrees to the normal. The main dimensions of the hole were $\frac{1}{2} \times \frac{5}{8}$ inch.

GERMAN LARGE HOLLOW CHARGE RIFLE GRENADE

(Gross Gewehr Panzergranate)

(Fig. 4)

The grenade is fired from the 3 cm. rifled discharger cup which can be fitted to the 7·92 mm. Mauser rifle and also to a modified version of the 7·92/13 mm. A-tk. rifle (Pz.B.39). The propellant cartridge contains a wooden bullet.

The rustproofed steel body of the grenade tapers slightly towards the head (1·8 to 1·5 inches) and is fitted with a parabolic impact cap, also of steel. At the base of the body a cylindrical tail unit is fitted. This is 2·9 inches in length and 1·2 inches in diameter with a pre-engraved 3 cm. band near its base. The tail unit may be of aluminium alloy or may be of black or brown plastic with a steel connecting shank and contains a graze fuze and a gaine. The overall length of the grenade is 7.15 inches and the weight, 13 oz. 5 drs.

The body contains a 4 oz. 6 dr. charge of cast T.N.T. in which the " hollow " is in the form of a conical cavity with a steel liner. The liner is coned at 19·5 degrees, has the apex cut away and the hole sealed by solder. At the top it is shaped to engage a supporting cannelure in the body and is fitted with a steel disc which has a small central hole. The impact cap is supported on the disc and the assembly is secured by the spun top of the body. At the base the body is reduced in diameter to match the tail unit and contains a 2·75 dram exploder pellet of P.E.T.N./wax. The base is closed by two cardboard washers.

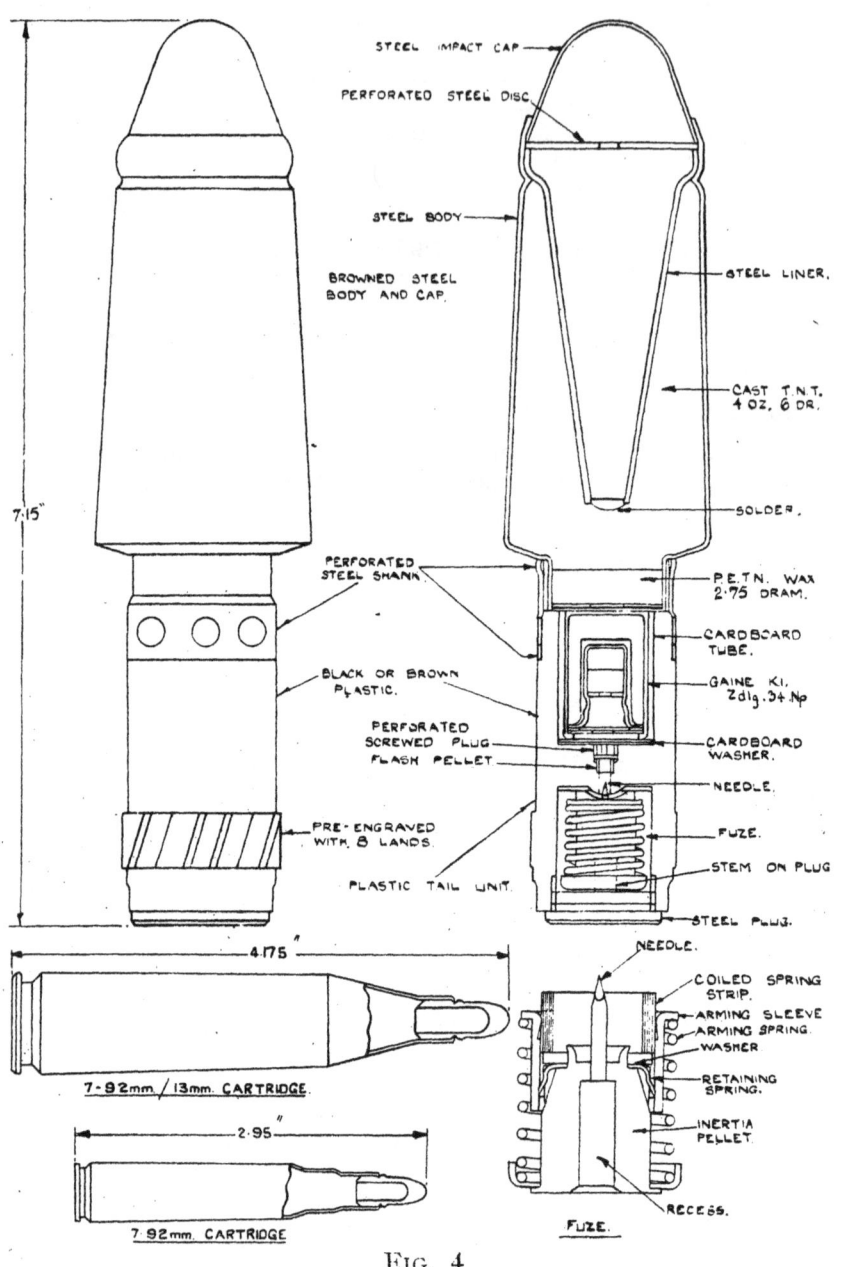

Fig. 4

The plastic tail unit is connected to the body by a steel shank which is screwed to the body and secured to the plastic by the latter being extruded into perforations in the shank. Eight rifling grooves are formed in the band near the base of the unit. The interior is recessed from the top to accommodate the small cylindrical gaine and from the base to receive the fuze with its steel screwed base plug. The diaphragm between the upper and lower recess is drilled to provide a passage for the needle and to receive a flash pellet which is secured by a perforated screwed plug.

The gaine, Kl.Zdlg. C/34, is described as a separate item in this pamphlet.

Fuze

The fuze consists of an inertia pellet carrying a needle which is held away from the flash pellet by a coiled spring strip supported on a washer secured to the head of the pellet. The coil is held within an arming sleeve which is supported by a spiral spring held in compression between an external flange on the sleeve and a collar at the base of the inertia pellet. The sleeve is prevented from rising, under the pressure of the spring, by a retaining spring which is secured to the head of the pellet and has four arms which engage a circular groove inside the sleeve. A second internal groove is formed near the top of the sleeve to engage with the arms when the fuze is armed.

Cartridges

The 7·92 mm. and 7·92/13 mm. S.A. cases are of varnished steel and are fitted with short hollow bullets of wood. The 7 92 mm. case is extended in length at the mouth and is coned into a cannelure in the bullet. The propellant charge consists of 19·6 grains of nitrocellulose powder in the form of square flake with a ridge across the centre of one side of the flake. The flakes are ·055 inch square, ·01 inch thick and the ridge protrudes ·01 inch. The ridge is probably intended to prevent the flakes packing tightly together. The composition of the propellant includes 98·36 per cent. of nitrocellulose (nitrogen content 13·12 per cent.) and ·88 per cent. of diphenylamine. The 7·92/13 mm. propellant charge consists of 51 grains of nitrocellulose powder in the form of granular tubes. The dimensions of the tubes in inches are: length ·050, external diameter ·084 internal diameter ·04.

Action

On acceleration the arming sleeve sets back, compressing its spring, and is prevented from rebounding by the arms of the retaining spring entering the internal groove near its head. The coiled steel strip is then free to expand so that on graze the pellet passes through the coil and the needle pierces the flash pellet. The flash enters the detonator in the gaine which brings about the detonation of the bursting charge with the assistance of the exploder. The functioning of the fuze, as the result of the strike of the impact cap, brings

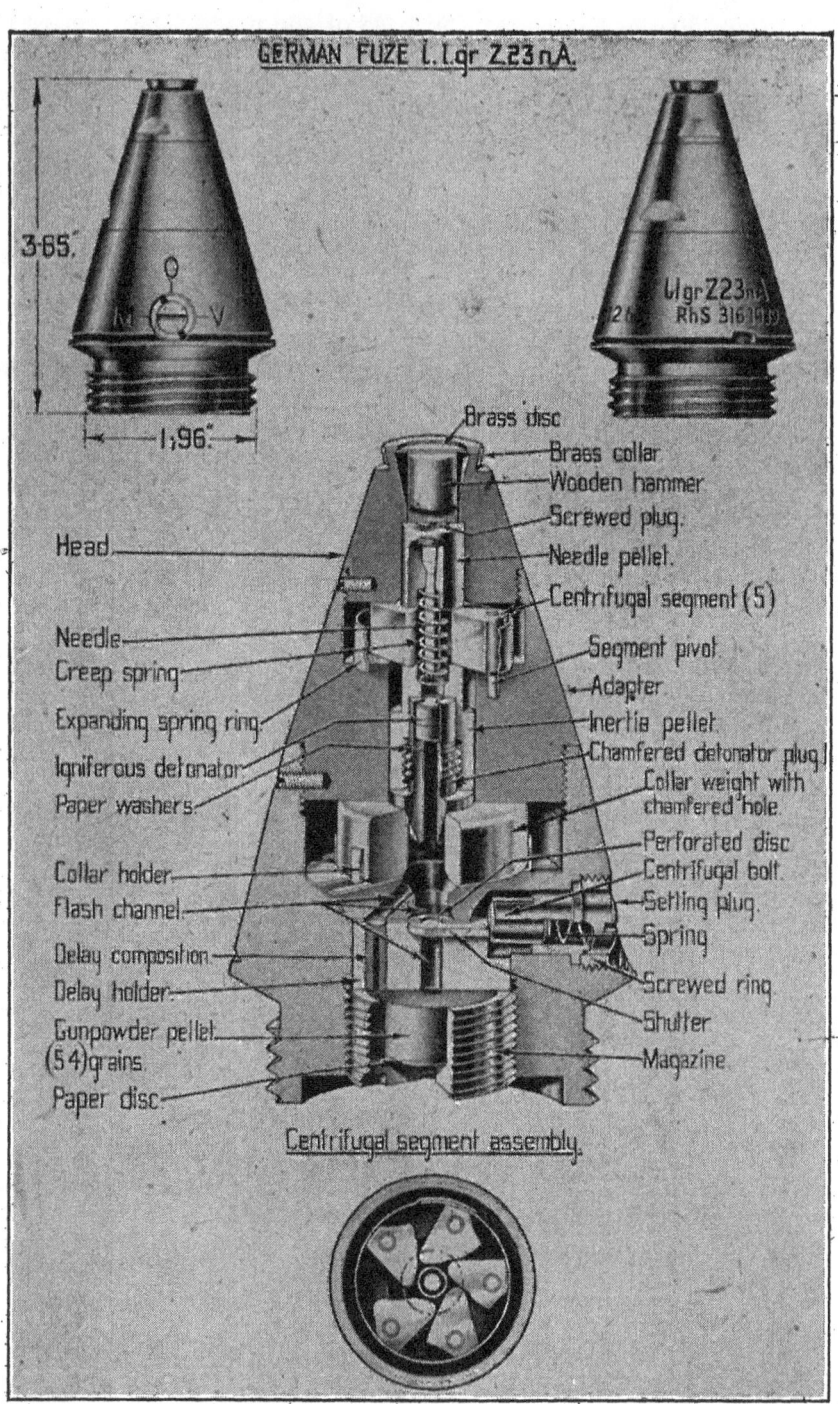

Fig. 5

about the detonation at the effective point relative to the surface struck and the hollow in the head of the charge results in the concentration of the effect of the detonation to obtain perforation. The steel disc under the cap is probably intended to prevent debris from the cap entering the liner.

Perforation of Armour Plate

Perforation of 50 mm. homogeneous hard armour has been obtained at 30 degrees to the normal. The main dimensions of the hole were $\frac{5}{8} \times \frac{3}{8}$ inch.

GERMAN FUZE 1.Igr. Z.23nA

(Fig. 5)

The fuze is used in the H.E. shell (Igr. 18) and the H.E./A1 shell (Igr. 18 Al) for the 7·5 cm. light infantry howitzer 1.I.G.18. A brief description of the fuze, under the designation " J.Gr.Z.23nA ", is included in Pamphlet No. 1. References are also made to it in the description of the 7·5 cm. light infantry shell in Pamphlet No. 8.

As will be seen by the drawing, the fuze is of the normal 23 type with an optional delay of ·15 of a second, but has a collar weight in the form of a cylindrical iron ring below the graze pellet. The collar has a chamfered central hole to engage a corresponding surface on the detonator plug which protrudes from the base of the graze pellet. The collar is retained in a central position within the fuze by a holder which consists of a brass washer with arms cut and bent up to engage the wall of the collar. A lateral thrust on graze would cause the collar to move sideways, and an action similar to that of an " Allways " fuze would be brought about by the inclined surfaces of the collar and detonator plug. The weight of this fuze is $12\frac{1}{2}$ oz.

GERMAN FUZE A.Z. 35 K.

(Fig. 6)

The fuze is of the igniferous type with a combined graze and direct action mechanism and an optional delay setting device which produces a ·3 second delay. In conjunction with a gaine it is used in the H.E. shell, " K.Gr.39 " in the 17 cm.K. in Mrs. L. equipment (a 17·25 cm. gun on the 21 cm. howitzer mounting).

Constructed mainly of steel, the portion protruding from the shell is coned with a flat top and is closed at the nose by a brass disc. The designation, " A.Z. 35 K." is stamped near the base of the coned head and, diametrically opposite, is a delay setting plug with setting marks lettered " M " and " O " stamped adjacent to

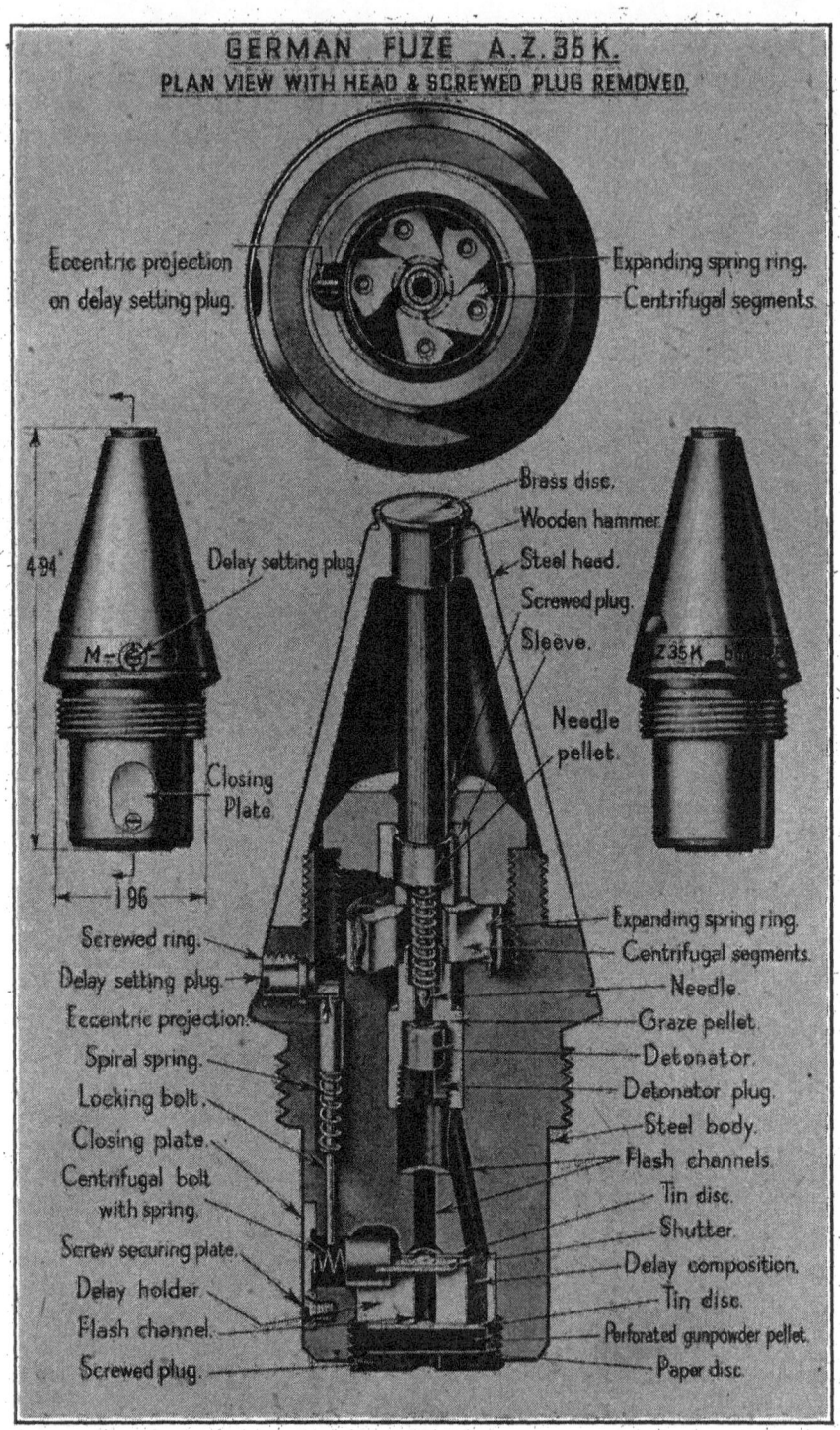

Fig. 6

it on either side. The overall length of the fuze is 4·94 inches and the weight 2 lb. 3 oz. 5 dr. The portion which is exposed when in the shell protrudes to a length of 3 inches.

The fuze body is screwthreaded externally for insertion in the shell and below the thread it is reduced in diameter and has a side aperture for the assembly of a centrifugal bolt attached to a shutter. The aperture is closed by a plate secured by a screw. Above the threaded portion a flange is formed which is coned to match the head. Above the flange the body is reduced in diameter and screwthreaded externally to receive the head. The delay setting plug, engraved with an arrowhead, is secured in a radial hole in the flange and on its inner end has an eccentric projection which bears on the upper end of the shutter locking bolt. The body is recessed from the base to accommodate the delay arrangement with a perforated pellet of gunpowder and is closed by a screwed plug with a central flash hole. Two flash channels in the top of the recess communicate with an upper recess containing a steel graze pellet carrying the detonator. This recess is enlarged near the top to form a platform for five pivoted centrifugal segments of aluminium. The segments are held in a position to overlap a shoulder on the graze pellet by an expanding spring ring surrounding them. The needle, carried in a cylindrical pellet of aluminium, is supported above the graze pellet by a spiral creep spring, the pellet being kept in alignment with the wooden hammer in the fuze head by a steel sleeve which fits over and protrudes above the pellet. The pellet and sleeve are accommodated in a central hole in a screwed plug which closes the top of the upper recess.

The delay holder consists of an aluminium cylindrical pellet with a flash channel through the centre, and a second channel, displaced from the centre, which contains a pressing of delay composition. A guideway for the shutter is cut in the top of the pellet and extends beyond the central open flash hole. The figures " 0, 30 " (indicating a delay of ·3 seconds) are stamped in the base of the holder. A thin disc of tin placed over the top of the holder has two perforations which correspond with the holes in the holder and is cut away to permit movement of the centrifugal bolt.

The shutter consists of a copper plate which fits in the guideway on the delay holder and is connected at its outer end to a centrifugal bolt. The bolt is pressed towards the centre of the body by a spiral spring held between the outer end of the bolt and a ring shaped recess in the plate closing the aperture in the side of the body. The shutter locking bolt is contained in a vertical channel between the top of the aperture and the top of the body and consists of a solid cylindrical pellet of steel with a stem at its base. A spiral spring surrounding the stem supports the pellet, the upper end of which is engaged by the eccentric projection on the inner end of the delay setting plug.

Action

When delayed action is required the delay setting plug is turned so that the arrowhead engraved on its head is set to the "M" graduation on the flange. At this setting the eccentric on its inner end is lowered and the shutter locking bolt is pushed, against its spring, down into the aperture in the side of the fuze body, thus preventing the centrifugal bolt from moving outwards. For non-delay action the plug is set to the "O" graduation. The eccentric is then in the raised position and the shutter locking bolt is clear of the path of the centrifugal bolt.

During flight the coil of the expanding spring ring is enlarged and the segments swung clear of the graze pellet by centrifugal force. When set for non-delay the centrifugal bolt is also thrown outwards, taking with it the shutter and thus exposing the open flash hole in the centre of the delay holder. If set for delay the centrifugal bolt is held by the locking bolt and only the channel containing the delay composition is exposed. During the period of deceleration, "creep" of the graze pellet is prevented by the creep spring. On graze the pellet, overcoming the spring by its momentum, moves forward and the detonator is pierced by the needle. When suitable impact is obtained, the hammer, and consequently the needle, are driven in at the same time as the pellet moves forward and direct action results. The path of the flash from the detonator to the magazine is governed by the setting. If set for delay, the empty channel in the delay holder is masked by the shutter and the path is through the channel containing the delay composition. If set for non delay the flash passing through the exposed empty channel will reach the magazine first.

GERMAN MECHANICAL TIME AND PERCUSSION FUZE S/90/45
(Dopp. Z.S/90/45)
(Figs. 7, 8 and 9)

The fuze has a mechanical time action with a maximum time of running of 90 seconds and a graze action of the normal German type. The time mechanism appears to be a 45 second mechanism modified by a change in the ratio of the gear train and an increase in the strength of the main spring. The fuze is possibly used as an alternative in the H.E. shell (K.Gr. 39) and the H.E.B.C. shell (K.Gr.38 (Hb)) for the 17 cm. K. in Mrs.Laf. The weight of the fuze is 1 lb. 12 oz.

In external appearance the fuze consists of a flat topped cone-shaped head of aluminium secured to a dull steel body near its base by a bright alloy securing ring. The upper part of the body is coned above the flange to match the head. Below the flange it is screwthreaded for insertion in the shell and below the thread there

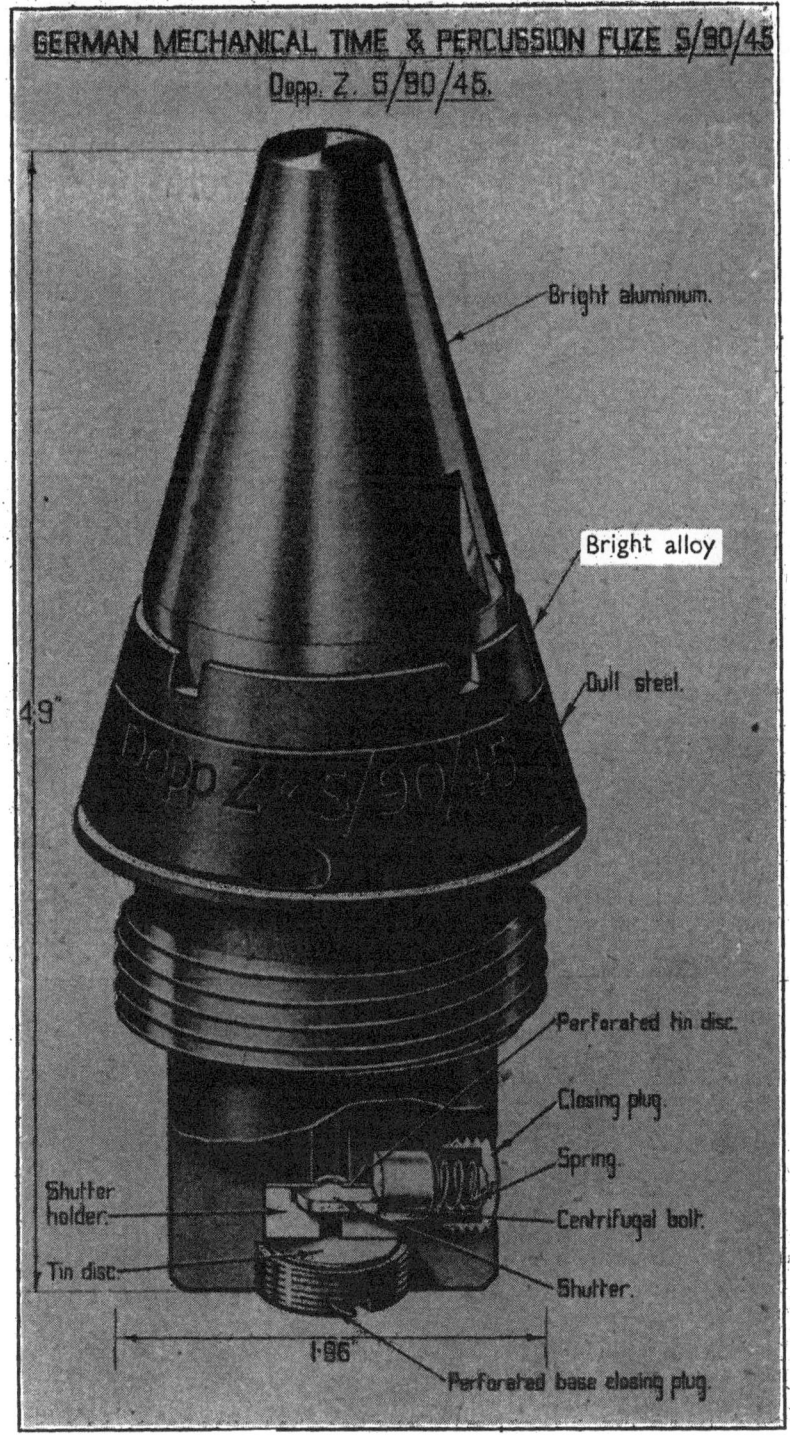

Fig. 7

is a plain cylindrical portion. The designation of the fuze, " Dopp Z.S/90/45 ", is stamped in the body above the flange. Rectangular key slots for setting are formed in the head and the body and a setting arrow for percussion action is engraved in the head adjacent to the slot. When inserted in the shell the fuze protrudes to the extent of approximately 3 inches. The diameter of the screw-threaded part is 1·96 inches.

Body

The body is recessed at the base to accommodate the shutter assembly and, equally spaced around the recess, there are three vertical channels for the bolts securing the time mechanism unit. An additional vertical channel, closed at the base by a screwed plug, accommodates the holder of the detonator for the time action. A radial channel in the wall of the recess contains a brass centrifugal bolt and is closed at the outer end by a screwed plug. A portion of the top of the recess is cut away to permit movement of the bolt and a flash channel in the centre of the top connects with a recess in the upper part of the body.

The upper recess contains the graze pellet of the percussion mechanism and is connected to the channel containing the detonator for the time action by an inclined channel which enters the recess near its base. Near the top, the recess is enlarged to form a platform on which four centrifugal segments of brass are fitted on steel pivots. Above this the recess is screwthreaded to receive the percussion needle holder. The coned flanged top of the body is recessed to form a platform for the time mechanism unit and is screwthreaded internally to receive the securing ring for the aluminium head. Two holes, diametrically opposite, are formed in the platform to receive the locking plungers of the time unit. There is also a locating stud for the unit and a circular seating for a press-pahn washer. Four small screws are inserted through the coned flange to lock the securing ring when the fuze is tensioned during assembly.

Percussion Mechanism

The graze pellet, carrying an igniferous detonator secured by a perforated screwed plug, is of brass with a shoulder near the top which is engaged by the four centrifugal brass segments. The segments are retained in a position overlapping the pellet by an expanding spring ring. A spiral creep spring is fitted between the pellet and a needle holder which consists of a steel disc screwed into the top of the recess containing the graze mechanism. The needle protruding from its underside has a pyramid shaped point.

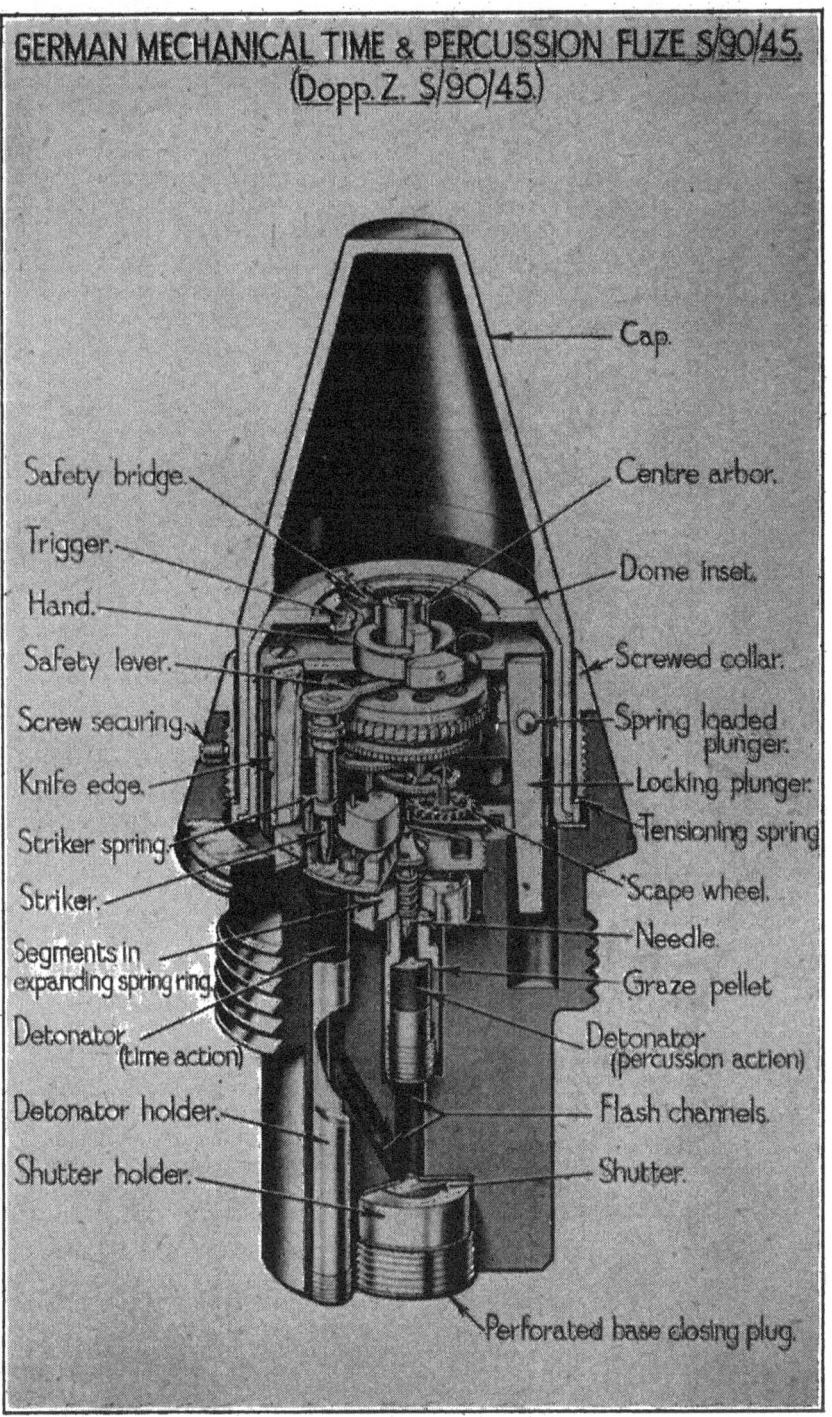

Fig. 8

Shutter Assembly

The assembly contained in the lower recess in the body consists of a copper plate shutter moving in a guideway formed in the top of a cylindrical aluminium holder. The holder has a central flash hole and is recessed at one side for the centrifugal bolt which, under the pressure of its spiral spring, bears against the outer end of the shutter and keeps the shutter in a position masking the flash hole in its aluminium holder. The centrifugal bolt is a cylindrical pellet of brass with a recess at its outer end to locate the inner end of its spiral spring. The outer end of the spring is held under compression by the screwed plug closing the radial channel in the wall of the recess and located by a circular groove in the inner face of the plug. A thin disc of tin with a central flash hole and with part of its circumference cut away for the centrifugal bolt is placed on top of the shutter holder. The recess containing the shutter assembly is closed at the base by a screwed plug which has a central flash hole. The flash hole is closed by a thin disc of tin fitted between the inner face of the plug and the base of the shutter holder.

Head Assembly

The flat topped aluminium head is in the shape of a cone with a cylindrical portion near the base. An external flange is formed at the base which supports the tensioning wire. The underside of the flange is cut away at four places to receive corresponding projections at the base of the dome. The dome, fitted inside the head, is of aluminium and is in the form of an inverted cup which fits over the time mechanism unit and is keyed to the flange at the base of the aluminium head by four projections at its base. The inside of the top of the dome is shaped to form a hand race against which the hand on the top of the time unit bears, when rotating. Part of the race is cut away so that the hand can be pushed upwards by its spring when in alignment with the slot so formed.

The head of the fuze is secured to the body by a securing ring which screws into the internal screwthread above the flange. Between the base of the securing ring and the flange at the base of the head there is a length of waved spring wire which is compressed between the ring and the flange when the ring is screwed down. By this means the tensioning of the head is adjusted during assembly so that it can be turned for setting but will not slip.

Time Mechanism Unit

The time mechanism is assembled in a cylindrical unit of superimposed brass plates with an aluminium hand on the top and two locking plungers in slots cut down diametrically opposite sides of the cylindrical assembly. The plungers are comparatively weighty

GERMAN MECHANICAL TIME & PERCUSSION FUZE S/90/45
(Dopp Z. S/90/45)
DIAGRAMMATIC ARRANGEMENT OF MECHANISM.

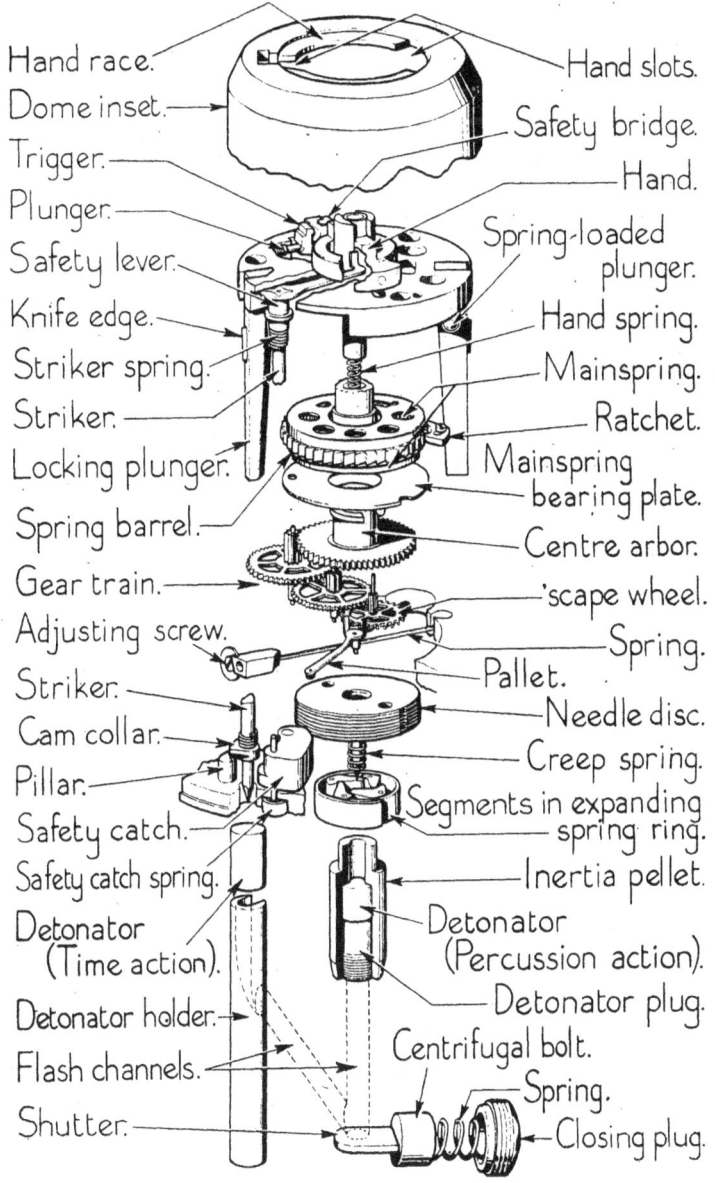

Fig. 9

and consist of steel strips tapering towards the base on the inner side. On the outer side of each, approximately at the centre, a projecting vertical knife edge is formed. A spring loader ball protruding from one side of the plungers engages a corresponding recess in the slots and thus supports the plungers which protrude below the base of the unit into recesses in the fuze body.

The mechanism is similar to that in the S/30 and S/60s fuzes, that is, it consists principally of a spring loaded striker held off the detonator by an eccentric cam collar on the striker which is supported by a pillar and a centrifugal safety catch. The cam collar is kept in this position by a safety lever which fits over flats formed on the forward end of the striker and is held by the ring shaped centre of the hand. The hand, with a spiral spring beneath it, is keyed to rotate with the centre arbor under the control of a train of gear wheels and an escapement and, before firing, is held by the pivoted trigger. Adjacent to the trigger a safety bridge is fitted which overlaps the hand to prevent functioning at settings of less than approximately one second. The spring plunger supporting the trigger engages in the end of trigger instead of the outer side, as with the S/30 and S/60s fuzes. The use of a stronger main spring, which is of wider strip than the original, has resulted in the use of a thin steel disc instead of the normal base of the spring barrel to support the spring.

Time Action

The time of running is governed by the size of the arc extending counter-clockwise between the position of the hand when held by the trigger and the position of the slot in the dome hand race. The fuze is set by turning the head with the aid of a setting device which consists of a graduated ring surrounding a moveable ring on which there is a setting index and a handle. The device is placed over the fuze, and the outer ring, bearing the graduations, is clamped by a key engaging in the key slot above the flange in the fuze body. A key in the inner ring, bearing the index, engages in the key slot in the head of the fuze. The inner ring is then rotated by means of its handle until the index is aligned with the required graduation and takes with it the head of the fuze. The dome inset, with the hand race formed in its upper part, is rotated with the head to the set position. The turning of the head is retarded by the waved wire tension spring between the securing ring and the flange at the base of the head.

On acceleration the two tapered locking plungers set back into the recess in the top of the body. As their wider portions with the protruding knife edges move back through the slots, the knife edges cut into the wall of the dome and thus prevent rotation of the dome relative to the mechanism unit. At the same time, the trigger hinging on its pivot sets back and releases the hand. The small spring loaded plunger in the trigger is then free to emerge and thus prevents the trigger rebounding. The release of the hand enables the main spring to rotate the centre arbor under the control of the escapement. The hand, rotating with the arbor, moves clear of the safety bridge and is pressed up against the hand race in the top of the dome by the spiral spring in the top of the arbor beneath the hand centre.

During flight, the safety catch is swung clear of the cam collar on the striker by centrifugal force and the striker is then supported only by the pillar. When the rotating hand reaches the slot in the hand race it is forced upwards by its spring and thus releases the safety lever keyed to the top of the striker. The striker, with the lever, is then rotated by the pressure of its spring combined with the effect of the inclined surface on the cam collar bearing on the pillar so that the collar moves clear of the pillar. The striker is driven away from the safety lever by its spring and pierces the detonator. The flash from the detonator, directed by the shape of the detonator holder, passes into the lower part of the recess containing the graze pellet and follows the same course as that described for the percussion action.

Percussion Action

When set for percussion the arrow on the head of the fuze coincides with the index line on the securing ring and the key slots in the head and body are in alignment. At this setting, the slot in the hand race of the dome is masked by the safety bridge, so that although the hand is released when acceleration occurs, it cannot rise through the slot to release the safety lever and the striker.

During flight the coil of the expanding spring ring, surrounding the segments engaging the shoulder of the graze pellet, is enlarged and the segments are swung clear of the pellet by centrifugal force. At the same time the centrifugal bolt is thrown outwards, compressing its spring, and the shutter opens.

On graze, the graze pellet overcomes the creep spring by its momentum and carries the detonator forward to be pierced by the needle. The flash from the detonator passes through the hole in the base of the recess, through the channel exposed by the open shutter, perforates the tin disc and emerges through the hole in the screwed plug at the base of the fuze.

GERMAN FUZE A.Z.2492 FOR SPIGOT RIFLE GRENADE
(Fig. 10)

The fuze is used with the H.E., anti-tank, hollow charge, rifle grenade, where it also serves as a means of uniting the vaned tail unit to the streamlined body.

The details and action of the fuze are included in the description of the grenade in Pamphlet No. 6.

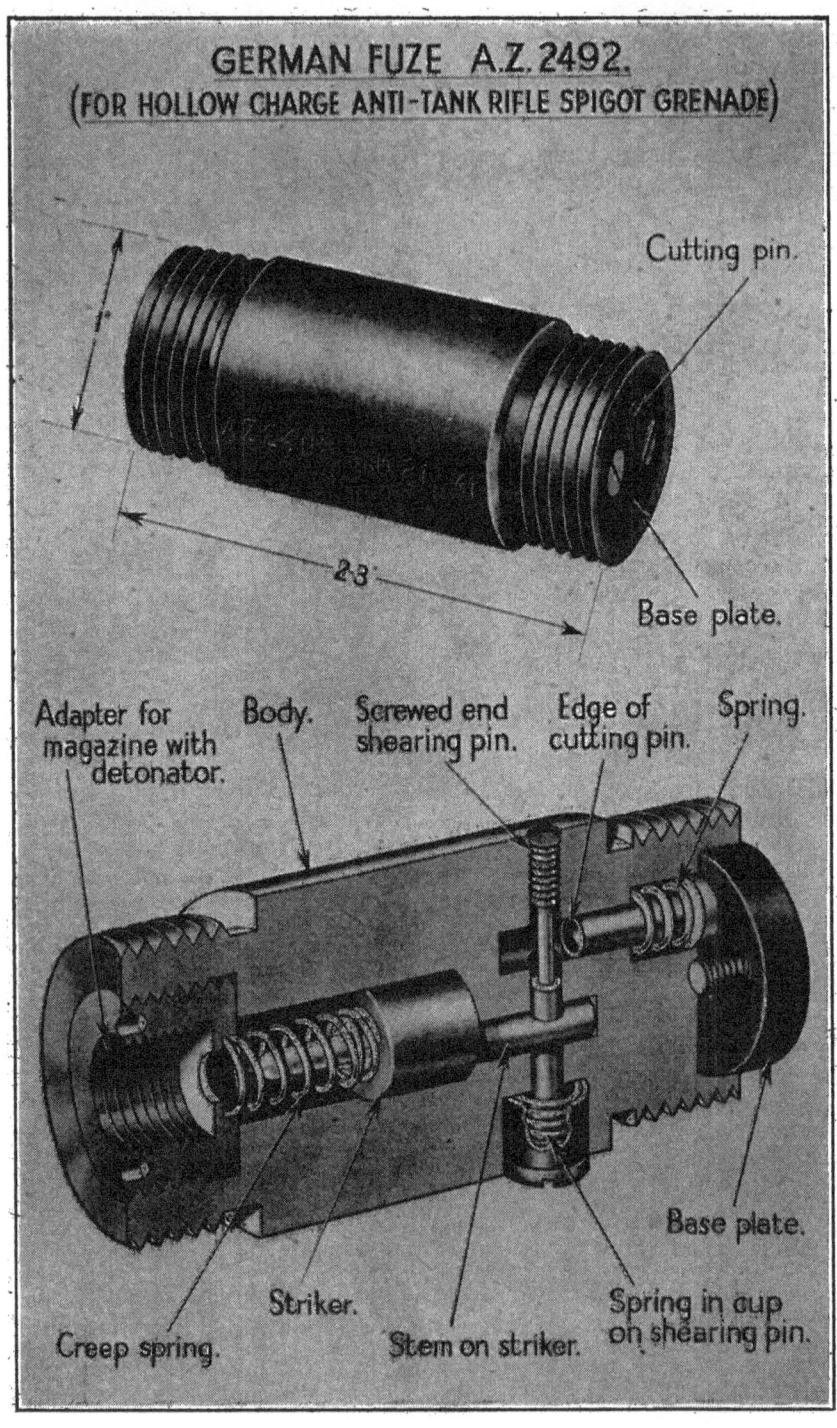

Fig. 10

GERMAN FUZE A.Z. 5072
(Fig. 11)

This small direct action detonating nose fuze has been found in H.E. shell for the 2·8 cm. Pz.B.41 and 4·2 cm. Pak 41 tapered bore anti-tank guns.

The aluminium ogival body and flat topped head of the fuze protrudes approximately ·6 inch from the shell. The designation, " A.Z. 5072 " is stamped in the body above the flange.

The weight of the fuze is 5·4 drams.

The body of the fuze is recessed at the top to receive the striker assembly and screwthreaded internally for the insertion of the head. At the base it is recessed and screwthreaded for the insertion of the magazine. The diaphragm between the two recesses has a pole at the centre for the striker.

The aluminium head of the fuze is recessed from the underside to fit over the striker assembly and has a thin diaphragm at the top.

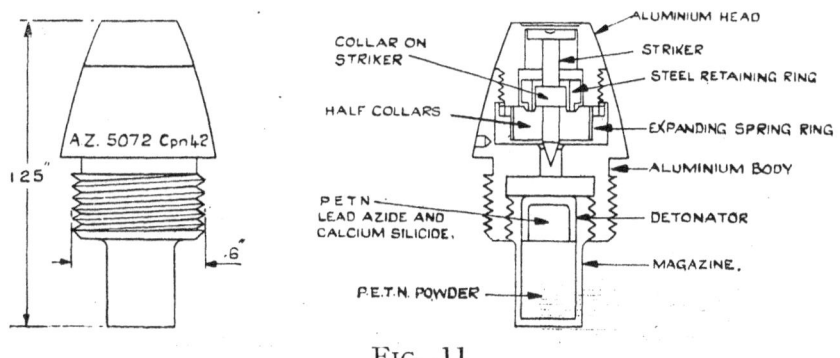

Fig. 11

The striker is supported by two centrifugal half collars which are retained in position under a collar on the striker by an expanding spring ring and a steel retaining ring. A circular projection on the base of the steel ring engages in corresponding semi-circular grooves in the top of the half collars.

The magazine contains P.E.T.N. in powder form with a detonator of lead azide, calcium silicide and P.E.T.N. at the top.

Action

During deceleration in flight the retaining ring moves forward and disengages from the groove in the half collars. The coil of the expanding spring ring is enlarged and the half collars thrown clear of the striker by centrifugal force. The striker is then held off the detonator by " creep " action. On impact the head of the fuze is crushed and the striker driven in to pierce the detonator.

GERMAN FUZE A.Z. 5075
(Fig. 12)

The fuze is used in the nose of the 3·7 Pak Stielgranate 41 or hollow charge muzzle stick bomb and, excepting the magazine and magazine assembly, is similar to the nose fuze described for the rifle and hand grenade in Pamphlet No. 6.

The aluminium body protruding from the nose of the bomb is ogival in shape with an aluminium striker protruding from a hole at the top. The weight of the fuze is 2 oz. 4½ drs. The designation has not been stamped on fuzes examined to date.

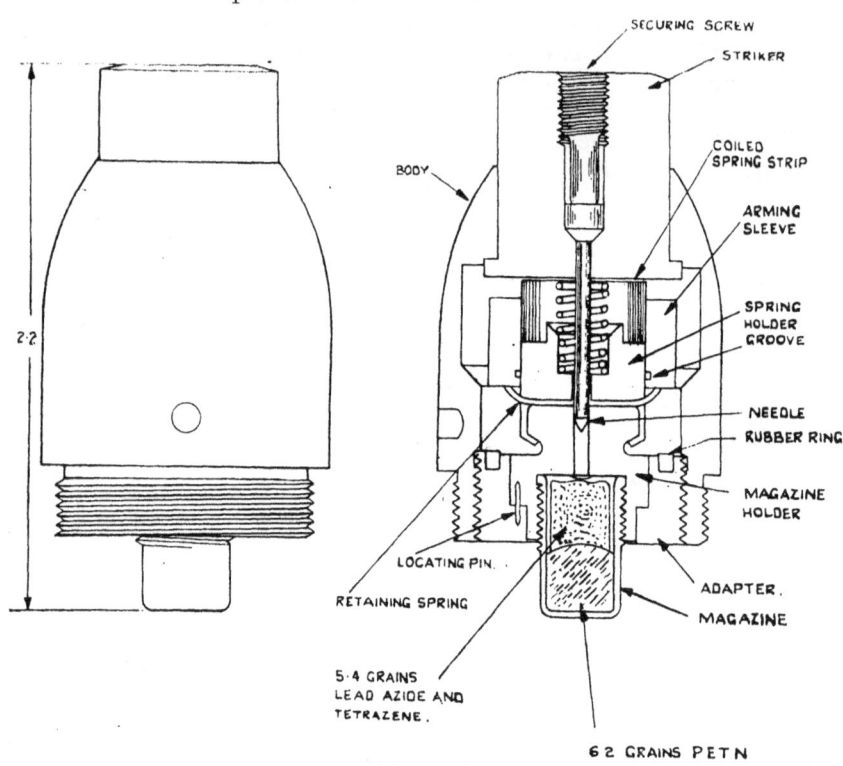

Fig. 12

The ogival aluminium body is screwthreaded externally for insertion in the bomb and is screwthreaded internally at the base to receive the adapter for the magazine. The interior is recessed for the mechanism, the recess being enlarged near the top to permit the expansion of a coiled spring strip.

The aluminium striker is cylindrical with a flange at the base which engages a step in the body and so limits its upward movement. A steel needle protruding from the base of the striker is secured by an aluminium screw inserted in the head. The striker is supported by a spiral spring and a coiled spring strip, both of which are positioned between the base of the striker and an aluminium spring holder.

The spring holder is in the form of a cylinder with a central hole for the needle and a recess in the top to receive the base end of the spiral spring surrounding the needle. Near the top the holder is reduced in diameter to form a step for the assembly of the coiled spring strip. The lower part of spring holder is recessed to fit over the top of the magazine holder. The wall of the recess is cut away at four places to permit the four arms of the retaining spring to protrude for the support of the arming sleeve. The retaining spring consisting of a steel disc with a hole for the needle in the centre and four arms curved slightly upwards, is held between the spring holder and the top of the magazine holder. The four portions remaining of the wall of the recess in the base of the spring holder are bent inwards to engage an inclined surface near the top of the magazine holder.

The steel arming sleeve, surrounding the coiled spring strip fits around the upper part of the spring holder and is supported by the retaining spring. Inside the sleeve, near the base, a circumferential groove is cut to engage the arms of the retaining spring when the sleeve is in the armed position.

The magazine holder consists of a cylindrical aluminium plug recessed and screwthreaded internally at the base for the insertion of the magazine. A shoulder formed around the exterior engages a corresponding shoulder in the adapter and carries a locating pin which enters a hole in the shoulder of the adapter. A projection at the top of the holder is bored centrally for the needle and is chamfered near its base for the attachment of the spring holder. The aluminium magazine contains 6·2 grains of P.E.T.N. under a 5·4 grain detonator. The detonator contains 93·7 per cent. of lead azide and 6·3 per cent. of tetrazene.

The aluminium adapter for the attachment of the magazine consists of a screwed ring with a stepped hole for the magazine holder. A rubber ring, to cushion the set back of the arming sleeve, is fixed in a groove in the top of the adapter.

Action

With the arming sleeve supported on the retaining spring, the coiled spring strip prevents the striker spring being compressed and thus holds the needle away from the detonator.

On acceleration the arming sleeve sets back on the rubber ring in the adapter and bends back the four arms of the retaining spring. Rebound of the sleeve is prevented by the arms entering the groove inside the sleeve. The coiled spring strip is then free to expand into the enlarged part of the body recess, leaving the striker supported only by the spiral spring.

On impact the striker is forced in and the needle pierces the detonator.

GERMAN FUZE BASE PERCUSSION NO. 5130
(Bd.Z.5130)
(Fig. 13)

The fuze is used in the 3·7 cm. Pak Stielgranate 41 hollow charge, muzzle stick bomb and is of the igniferous type with a graze action. The designation of the fuze, " Bd.Z.5130," is stamped in the base. The weight of the fuze is 4 oz. 12 dr.

The body of the fuze is cylindrical with a screwthreaded portion of reduced diameter at the head for insertion in the bomb. A large recess, formed from the head, contains the graze pellet and is closed at the top by the detonator holder. A smaller recess near the periphery is formed from the base and contains a detent supported by a spiral spring. This recess is closed at the base by a disc secured by the turned over metal of the body. A channel, with a downward incline towards the side, is drilled from the exterior of the body, through the recess containing the detent and into the recess containing the graze pellet. A ball which engages a shoulder on the graze pellet is held in this channel, between the two recesses, by the detent. Two key flats are formed near the base end of the body.

The graze pellet consists of a solid cylindrical pellet fitted with a needle at the head. An inclined shoulder, with which the ball engages, is formed below the head. A helical spring is held between the head of the pellet and the base of the detonator holder, the smaller end of the spring fitting around the needle.

The detonator holder consists of a flanged cup with a flash hole in its base and contains an igniferous detonator stamped with the number " 26 ". The detonator is retained by the mouth of the cup being turned in to overlap a washer. The holder is supported at its flange by a step formed in the top of the recess and is secured by turned-in metal of the body.

The detent is a solid cylindrical pellet with a concave head and a short stem at the base around which the upper end of spiral supporting spring is positioned.

Action

During transport the graze pellet is held away from the detonator by the ball, which is held in contact with the shoulder on the pellet by the detent.

On acceleration the detent, overcoming the spring, sets back and thus releases the ball, which moves down the inclined channel and leaves the graze pellet held off the detonator only by the helical

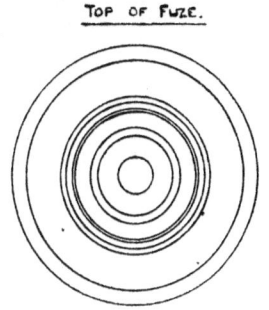
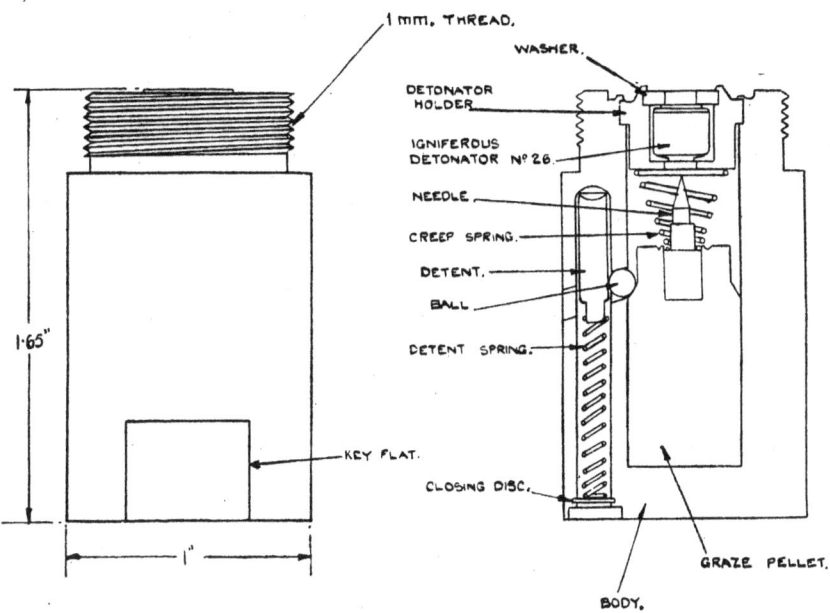
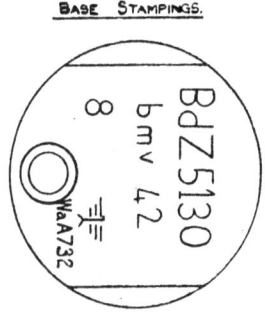

Fig. 13

creep spring. The concave head of the detent is apparently designed to receive the ball when the detent rises again, and so to push the ball up into the top of the recess.

On graze the pellet moves forward, compressing the creep spring and the needle pierces the detonator.

GERMAN SMALL P.E.T.N. GAINE NO. 34
(K1.Zdlg. 34 Np)
(Fig. 14)

The gaine is used in the hollow charge 3 cm. pre-engraved rifle grenades, the 5 cm. mortar H.E. bomb and the 3·7 cm. Pak muzzle stick bomb.

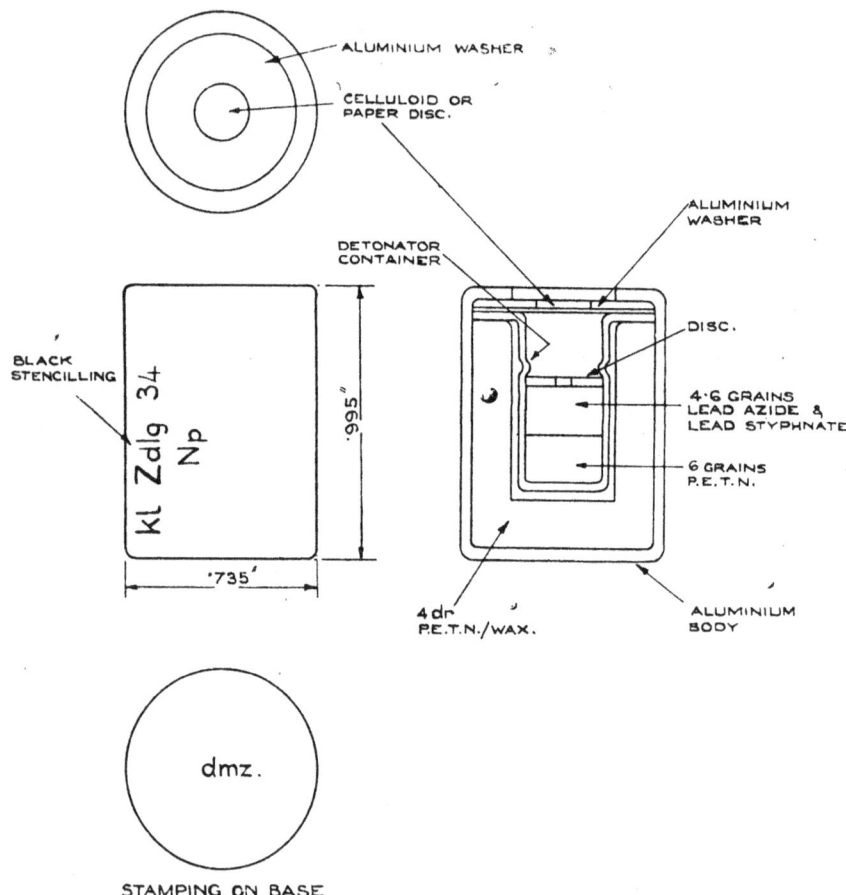

Fig. 14

The construction of the gaine is similar to that of the other types described in Pamphlet No. 6, that is, an aluminium cylindrical body with an aperture at one end which contains the detonator. The length is ·995 inch, the diameter ·735 inch, and the weight 6 drams. The exterior is stencilled " Kl.Zdlg. 34 Np ",

The main filling, as indicated by the marking " Np ", consists of P.E.T.N./wax and weighs approximately 4 drams. The detonator assembly, which closes the head end of the body and enters a cavity in the main filling, consists of a flanged container holding approximately 4·6 grains of lead azide and lead styphnate over 6 grains of P.E.T.N. This initiator filling is covered by a perforated disc. The container is closed at the top by a paper disc held between a cardboard washer and the overturned top of the body.

GERMAN SMALL P.E.T.N. GAINE 34 WITH DELAY
(Kl.Zdlg. 34 m.V.)
(Fig. 15)

The gaine is used in the 8 cm. mortar H.E. bombs, Models 38 and 39, and is distinguished from the Kl.Zdlg. 34 Np, also described in this pamphlet, by the letters " m.V." stencilled in black after the number on the exterior of the gaine.

The dimensions of the gaine are practically the same as the non-delay type, the small differences probably being within the tolerances permitted in manufacture. The essential difference is in the filling of the detonator container. The difference here, as shown in the drawing, consists of two additional compositions introduced above the cup containing the lead azide and lead styphnate and the use of another cup, with a flash hole closed by a fabric disc, as a means of closing the top of the detonator container.

The delaying effect obtained by the combination of the gaine and the delay chamber in the 8 cm. mortar bombs is reported to be approximately ·7 of a second.

GERMAN SMOKE BOX NO. 8 FOR H.E. SHELL
(Rauchentwickler Nr 8)

Some details of this smoke box were included in the description of the 7·5 cm. Kw.K. H.E. cartridge in Pamphlet No. 4, page 32.

The box consists of a bakelized cardboard cylindrical container 1·7 inches long and 1 inch in diameter, closed at the top and bottom by plastic discs. The composition weighs 28·3 grams and consists of :—

Red phosphorus	82·3 per cent.
Paraffin wax	12·8 per cent.
Magnesium phosphates	4·9 per cent.

In addition to the designation " Rauchentwickler Nr 8 ", the label on the top of the box examined bore the date 1941. The composition differs from that given in Pamphlet No. 4 by the substitution of magnesium phosphates for alumina.

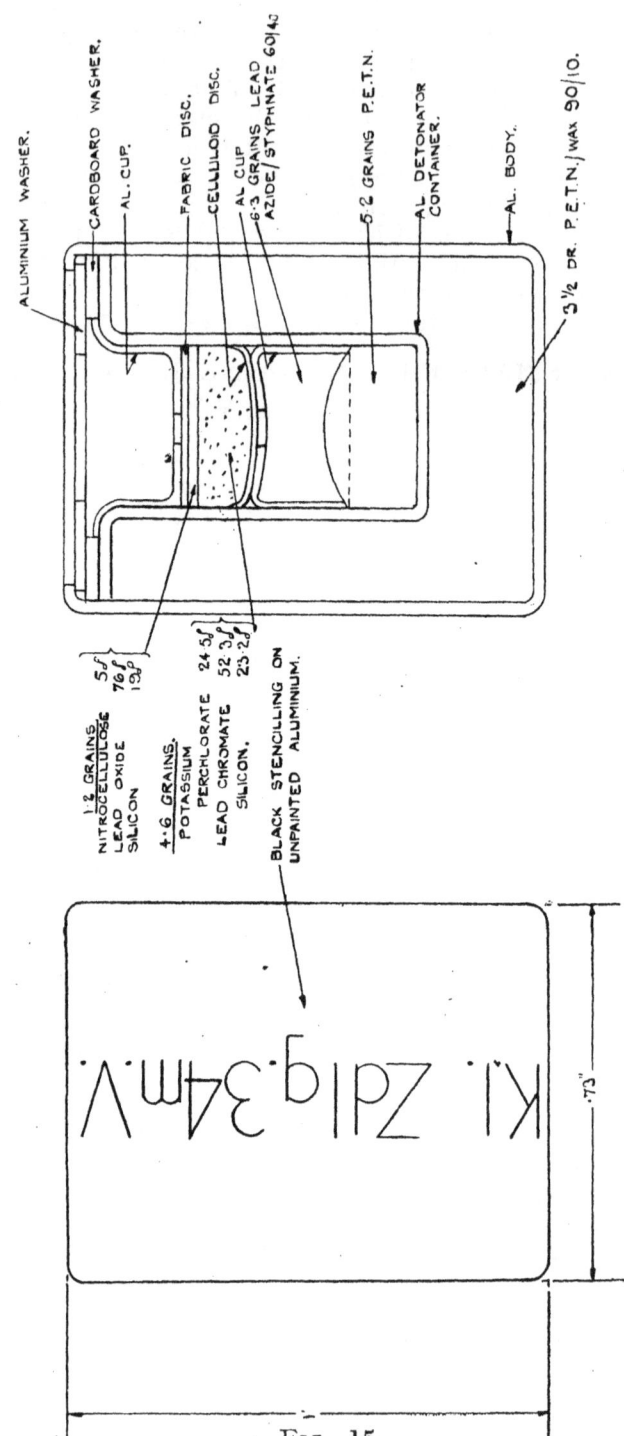

Fig. 15

GERMAN SMOKE BOX NO. 9 FOR H.E. SHELL
(Rauchentwickler Nr 9)

The box consists of a bakelized cardboard cylindrical container 2·3 inches long and 1·4 inches in diameter with a closing cap of plastic. The composition weighs 85·82 grams and consists of :—

Red phosphorus	82·3 per cent.
Paraffin wax	12·2 per cent.
Magnesium phosphates	5·5 per cent.

The box examined bore no markings.

GERMAN TRACER NO. 3
(Lichtspurhülse Nr 3)
(Fig. 16)

This type of tracer is used in 3·7 cm. Pak and Flak H.E. shell.

The tracer body of rustproofed steel is 1·77 inches long and ·9 inch in diameter with an external screwthread near the base for insertion in the shell. The front end is closed by a steel disc over which the rim of the body is turned. The base is closed by a celluloid disc. The filling is contained in a steel inverted cup, the interior of which is lubricated with graphite. The filling consists of two pressed pellets of tracing composition and a third base pellet of tracing composition with priming composition at its base. The pellets are covered with a thin even coating of graphite. Only the priming composition has a serrated finish. The priming is unusually small (a layer of about ·04 inch deep weighing 7·7 grains) and is pressed with the lower pellet.

The three increments of tracing composition total 197·5 grains and were found by analysis to consist of : magnesium metal 32 per cent., magnesium hydroxide 6·2 per cent., barium nitrate 36·3 per cent., barium carbonate 11 per cent., sodium oxalate 12·4 per cent., and 2·1 per cent. of organic matter in the form of a waxy substance.

The priming composition consists of : magnesium metal 27 per cent., barium nitrate 42 per cent., strontium nitrate 3 per cent., and 28 per cent. of organic matter which appears to be a phenolic resin.

The tracing composition spinning at 30,000 revolutions per minute burnt with a white light of 1,050 candle power. The estimated duration of the trace is 8 seconds.

The weight of the filled tracer is approximately 3 oz. 8 dr.

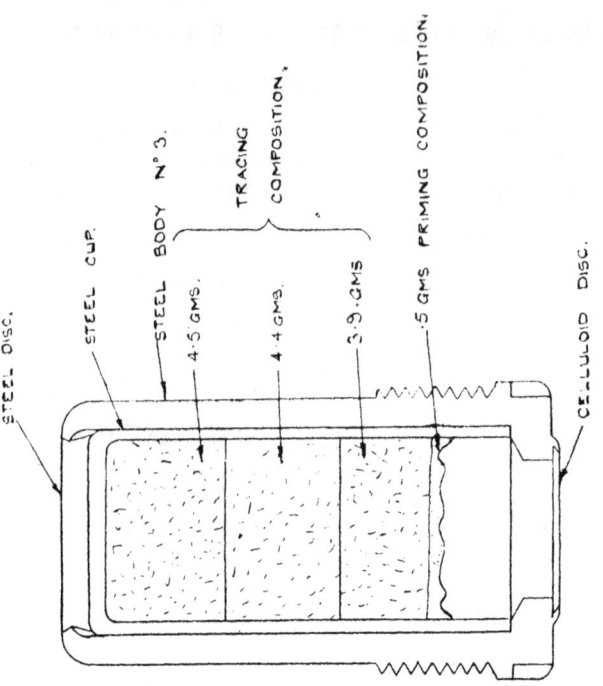

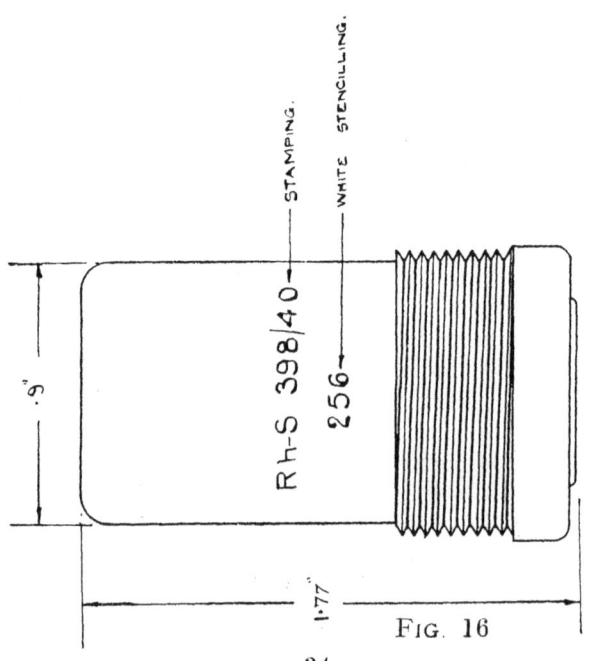

Fig. 16

GERMAN 2.7 cm. PISTOL H.E. CARTRIDGE
(Sprengpatrone für Kampfpistole)
(Fig. 17)

The cartridge is used with the 2·7 cm. " Kampfpistole ", a rifled pistol which also fires signal and smoke cartridges.

The complete round is 5·1 inches in length and consists of an unpainted aluminium projectile with five pre-engraved rifling grooves, a nose fuze of magnesium or browned steel with a protruding striker and a short rimmed case of aluminium with a brass bush in the base carrying a percussion cap. The red stencilling " Spreng Z " appears on the base.

The weight of the complete round is 5 oz. 2 dr.

Fuze and Gaine

The fuze is of the direct action type with an igniferous detonator and is fitted with a gaine at the base. The body of the fuze, which screws into the shell, is fitted with an ogival cover which forms the head of the projectile and from which the striker protrudes at the nose. The cover is provided with key flats and is secured to the body by a locking screw. The body is recessed and screwthreaded at the base for the insertion of the gaine, above which is a paper washer and a ring-shaped holder carrying the detonator. The upper part of the body is reduced in diameter to receive a retaining collar and to form a platform for three pins supporting the collar. A channel accommodates the striker with its supporting spring. The steel striker is retained in a cylindrical head by a screw inserted at the top of the head. The head is flanged near its lower end to retain it in the cover and is reduced in diameter below the flange to form a shoulder. Six steel balls are held between the shoulder and the top of the body by the retaining collar.

The gaine contains a 6·8 grain filling of P.E.T.N. with a 5·9 grain initiator of lead azide and lead styphnate.

Action of Fuze

On acceleration the supporting pins of the retaining collar collapse and the collar sets back clear of the balls. At this stage the balls are held by the set back of the striker head. During flight the balls are thrown clear by centrifugal force, leaving the striker supported only by the spring. On impact the striker is driven into the detonator. The flash from the detonator initiates the gaine, which detonates and thus brings about the detonation of the bursting charge.

The combination of igniferous detonator, the gaine and the gap below the gaine in the shell is probably intended to result in a slight delay. Steel fuzes and fuzes largely of magnesium have been met with. The magnesium fuzes may be intended to produce a flash on impact to assist observation or to obtain incendiary effect.

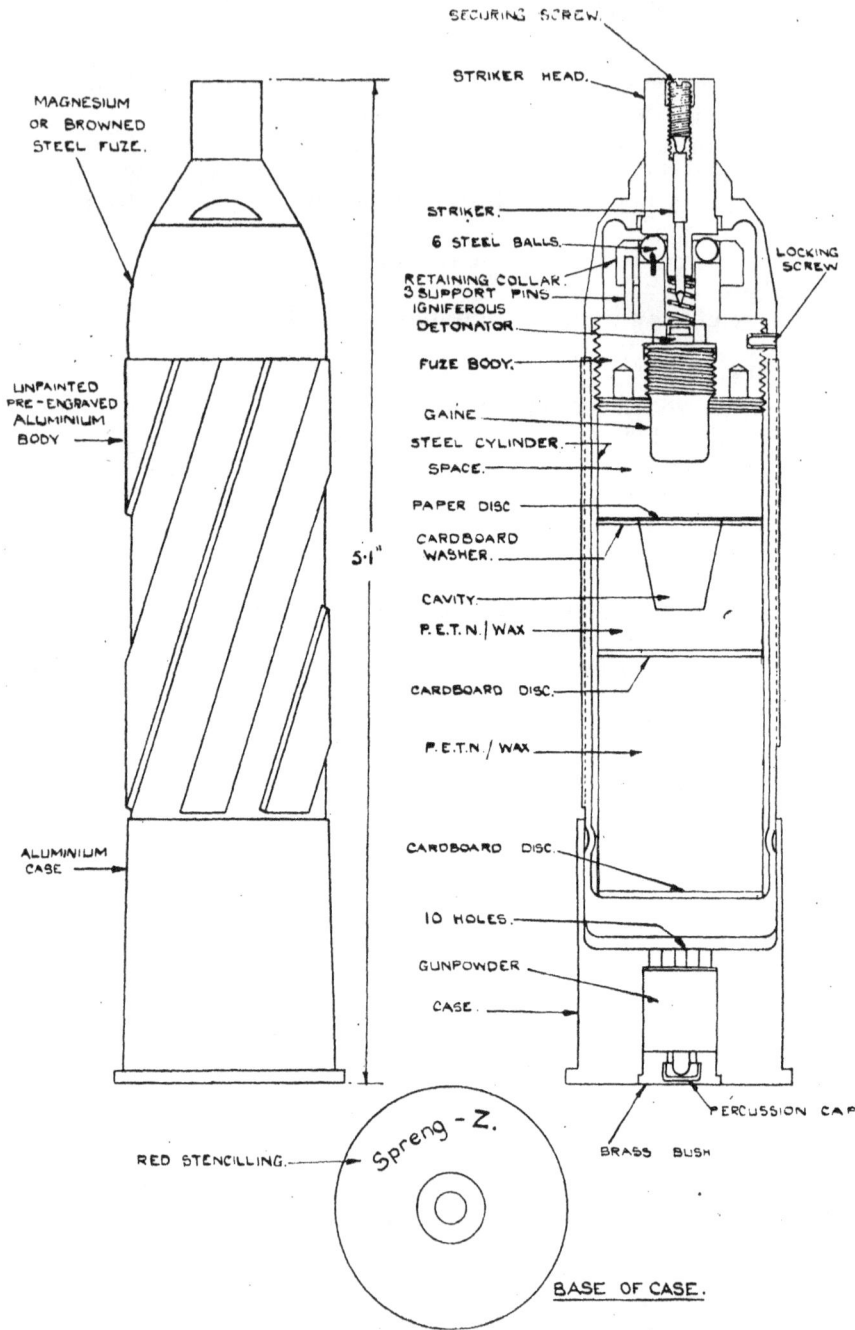

FIG. 17

Shell

The shell consists of an aluminium body, pre-engraved for most of its length at a right-hand twist of approximately 1 in 9 to correspond with the rifling in the pistol, and fitted with a steel liner which holds the bursting charge and is screwthreaded to receive the fuze. The liner is secured to the body by indentations and knurling near the base.

The bursting charge consists of approximately 340 grains of P.E.T.N./wax, 90/10, in two increments. A cardboard disc is inserted between the two increments and at the base of the lower increment. The upper increment is some distance below the gaine and has a tapering cavity. A cardboard washer carrying a thin paper disc is placed on the top of the increment.

The weight of the shell is approximately 3·5 oz.

Cartridge

The case, with its percussion cap carried in a brass bush, is the same as that described for the signal cartridge. The weight of the propellant charge of gunpowder is 12·3 grains.

GERMAN 2·7 cm. PISTOL SIGNAL CARTRIDGE
(Deutpatrone für Kampfpistole)
(Fig. 18)

The cartridge is used with the 2·7 cm. " Kampfpistole ", a rifled pistol which also fires H.E. and smoke cartridges. The projectile emits a puff of reddish brown smoke approximately 1·8 seconds after firing.

The complete round is 5·08 inches in length and consists of an unpainted aluminium projectile, with a tapering rounded head and five pre-engraved rifling grooves, which is secured to a short cartridge case of aluminium. A percussion cap is held in a brass bush in the base of the case. The base is stencilled " Deut Z ".

The weight of the complete round is approximately 4 oz. 6 dr.

Projectile

The projectile, weighing approximately 3 oz. 4 dr., consists of of an aluminium envelope containing the signal composition, pre-engraved for most of its length at a right-hand twist of approximately 1 in 9 to correspond with the rifling in the pistol. Two increments of the signal composition are pressed into the head, and below this is a light alloy cylinder which forms a liner to the envelope and contains eight further heavily pressed increments of signal composition. A hole through the centre of the cylinder filling contains three strands of quickmatch. Two additional strands are inserted laterally between the base of the filling and a light alloy distance piece below the filling. The distance piece is of light alloy and is in the form of a washer with a tubular projection at the base, in which there are nine holes. Below the distance piece the projectile is closed by a screwed base plug with a left-hand thread. The plug may be of light alloy or of plastic and has eight longitudinal holes arranged in a circle and closed at the base by a single cardboard washer. An additional hole in the centre of the plug is enlarged

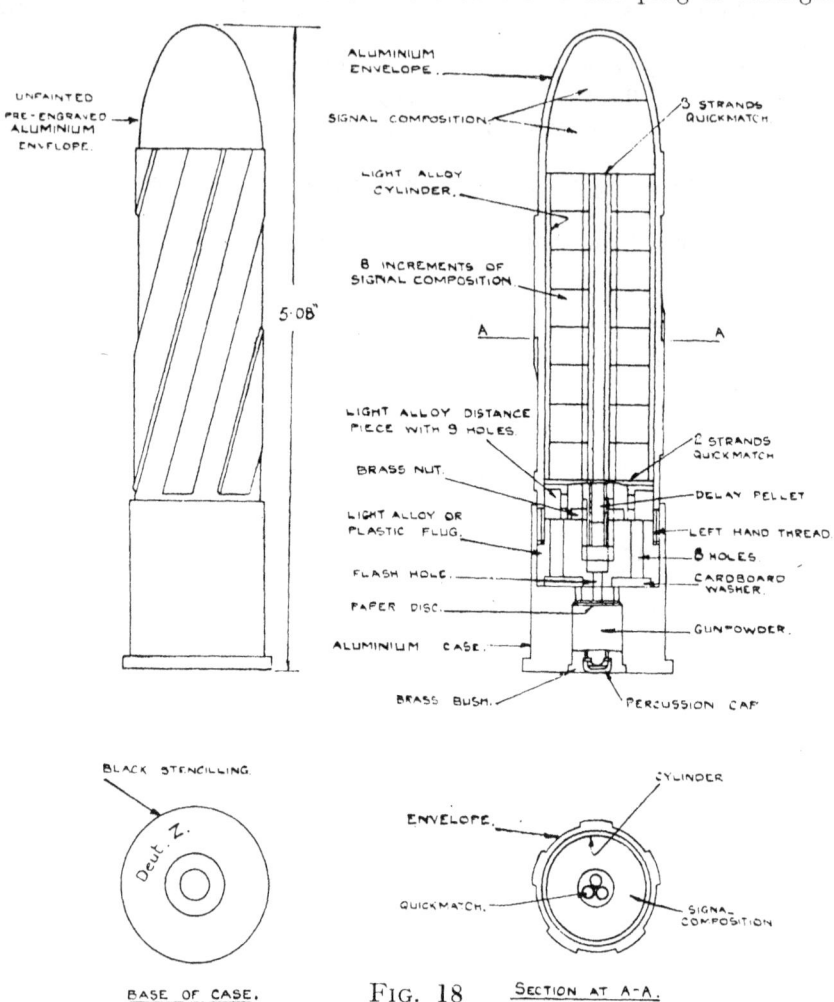

FIG. 18

and screwthreaded at the top to receive a delay pellet. The delay pellet consists of a brass tube threaded internally to hold a pressing of delay composition and externally for insertion in the plug and to receive a brass fixing nut which is also glued.

The signal composition consists of :—Sodium chlorate 25 per cent., a carbohydrate 38 per cent. and a dye 37 per cent.

Cartridge

The aluminium case is 1·33 inches in length and is rimmed. The propellant charge, consisting of 11·6 grains of gunpowder, is contained in a recess in the solid base from which nine flash holes lead into the upper portion where the projectile is secured. The holes are closed by a paper disc. The recess containing the propellant charge is closed at the base by a brass bush in which is carried a percussion cap. The bush is shaped to form a cap chamber with anvil and flash holes and is secured in the base by ringing.

Action

When the percussion cap is struck the flash passes through the holes to the propellant charge which ignites the delay composition through the central flash hole and at the same time projects the projectile by gas pressure through the other eight holes. After approximately 1·8 seconds the quickmatch is ignited by the delay composition and this, in turn, ignites the signal composition. The pressure set up blows the cardboard washer from the bottom of the base plug and a reddish brown smoke is emitted through the holes in the distance piece and thence through the holes in the base plug.

GERMAN 2·8 cm. Pz.B.41 CARTRIDGE, Q.F., H.E.

(Fig. 19)

This fixed Q.F. round is fired from the cone bored 2·8 cm. anti-tank gun Pz.B.41. The shell is fitted with a small nose fuze. Only the head and upper flange of the shell protrude from the neck of the case. The junction between the upper flange and the mouth of the case is painted red. The shell is painted black. The length of the complete round is 8·7 inches and the weight, 1 lb. 4 oz. 6 dr. The base of the case is stencilled " Sprgr Patr ", in white.

The complete round consists of the following components :—

    Fuze A.Z.5072.
    Flanged H.E. shell filled P.E.T.N./Wax.
    Brass case (without model number).
    Propellant charge of nitrocellulose powder in tubular granules.
    Percussion primer C/13nA St.

Fuze

A description of the fuze is included in this Pamphlet.

Shell

The steel body is apparently machined from bar with two integral sloping flanges. The upper flange, formed at the shoulder, is 1·11 inches in diameter and has five equally spaced holes of ·1 inch

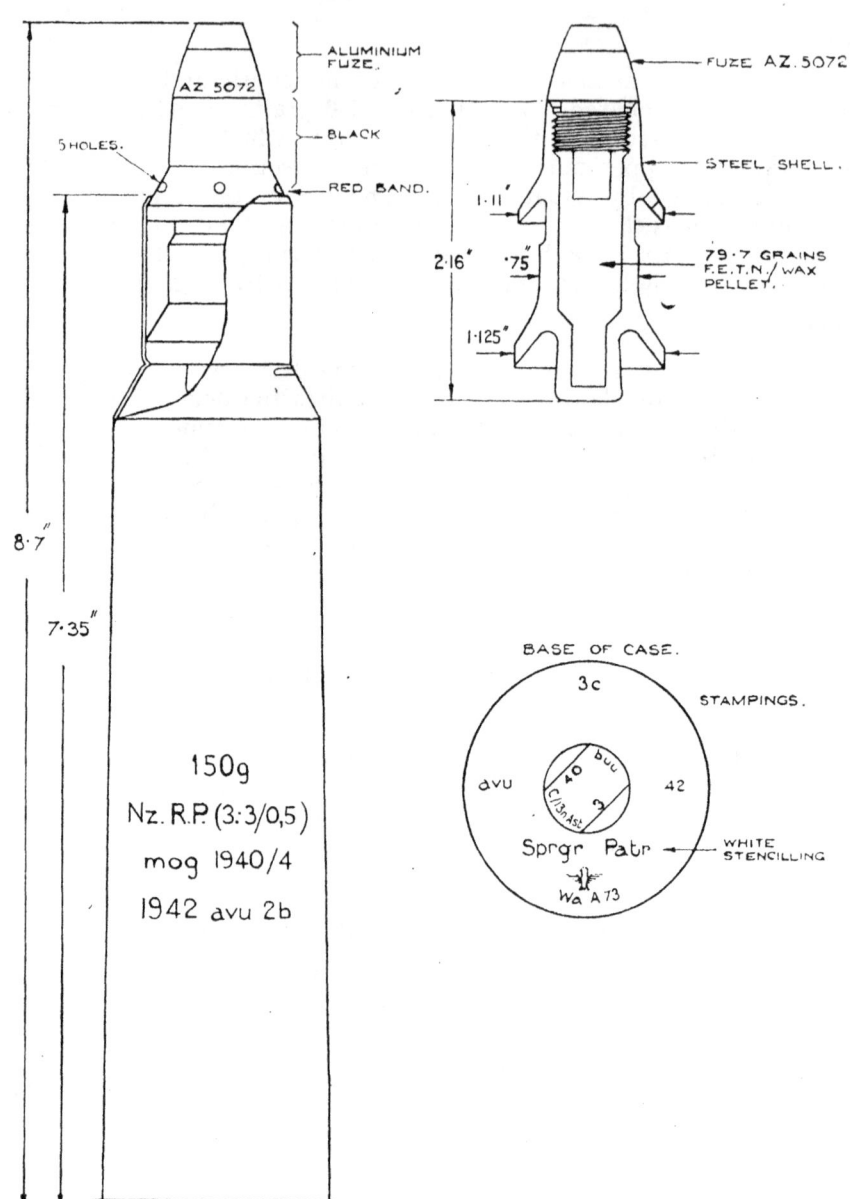

FIG. 19

diameter for the escape of air when the flange is pressed down during the travel up the bore. Below this flange the body is grooved to accommodate the pressed back flange. The lower flange is formed near the base and is 1·125 inches in diameter. Below the flange the body is reduced in diameter from ·75 inch to ·47 inch but increases to ·5 inch at the base. An internal screwthread for the fuze is formed at the head. The cavity for the bursting charge is cylindrical with a reduction in diameter in the lower part. The weight of the empty shell is 2 oz. 8 dr. The bursting charge consists of a 79·7 grain pressed pellet of P.E.T.N./Wax which, in accordance with the German practice, is dyed pink. The weight of the shell, filled and fuzed, is approximately 3 oz. (1,321·4 grains).

Case

The brass case is 8·7 inches in length, has the normal rim at the base and has a marked increase in taper below the neck. The shell is held in the necked portion of the case by three indentations below the base of the neck and by the mouth of the case being turned in to overlap the upper flange on the shell. The primer hole in the base of the case is screwthreaded for the insertion of the C/13 primer. Black stencilling on the side of the case indicates the weight, nature, shape, size, place of manufacture and year of the propellant, also the place and date of the filling of the cartridge.

Propellant Charge

The weight of the propellant charge, as indicated by the stencelling on the case varies, probably according to the proof results of the propellant lot. Weights varying between 137 and 150 grams have been met with. The propellant, " Nz.R.P. (3 · 3/0,5) " is a nitrocellulose powder in the form of short tubular grains. The dimensions of the grains, in inches, are: length ·118, external diameter ·118, internal diameter ·02.

Primer

The percussion primer C/13nA St differs from the C/13nA described in Pamphlet No. 7 in that it is made of steel instead of brass.

GERMAN 3·7 cm. Pak CARTRIDGE Q.F., H.E./TRACER
(3.7 cm. Sprgr. Patr 18 umg)
(Fig. 20)

This fixed Q.F. H.E. round, fired from the 3·7 cm. anti-tank gun, is similar to the round described in Pamphlet No. 6 except that the H.E. bursting charge is of T.N.T. This is indicated by the H.E. numeral " 27 " stencilled above the shoulder of the shell. The shell described in Pamphlet No. 6 is normally marked " 32 " to indicate the P.E.T.N./Wax 90/10 filling.

The shell, fitted with the A.Z.39 fuze, is painted aluminium colour and has a yellow band about midway between the single driving band and the fuze. This band distinguishes the H.E./T.

GERMAN 3·7 cm. H.E./T SHELL.
(SPRENGGRANATE 18 umg. FOR Pak AND Kw.K)
AZ39 cag42 24
Wa A647
TP 27
CE. 7. 7 41. 0
CE. 7. 7. 40. 0.
3·7"
BODY PAINTED ALUMINIUM
YELLOW BAND.
PAINTED BLACK.
LEATHER OR COMPOSITION WASHER
PAPER DISC.
ALUMINIUM CONTAINER.
14 DR PELLET PRESSED T.N.T FLAKE.
STEEL DISC.
STEEL CUP.
TRACING COMPOSITION.
PRIMING COMPOSITION.
TRACER BODY No 3
CELLULOID DISC.

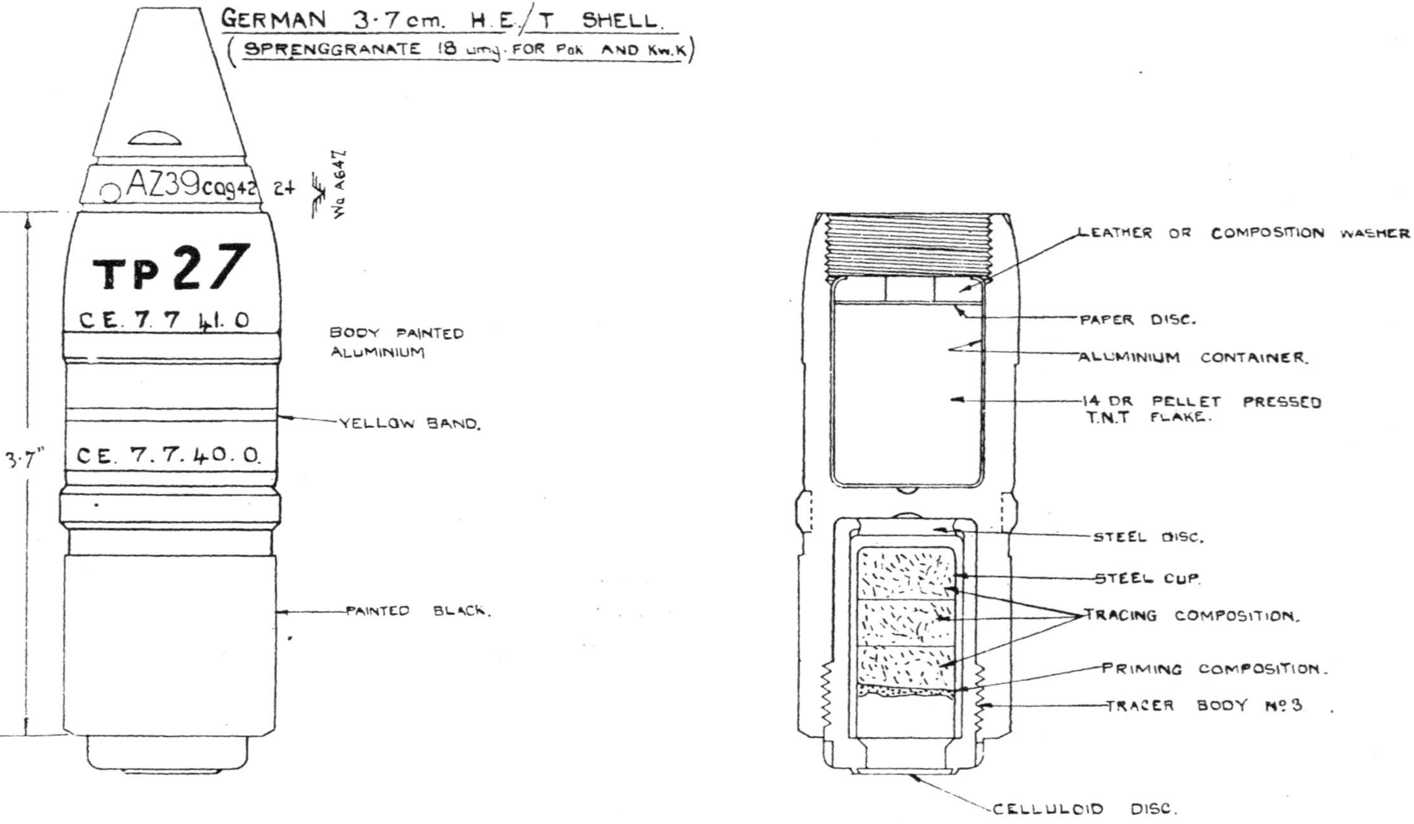

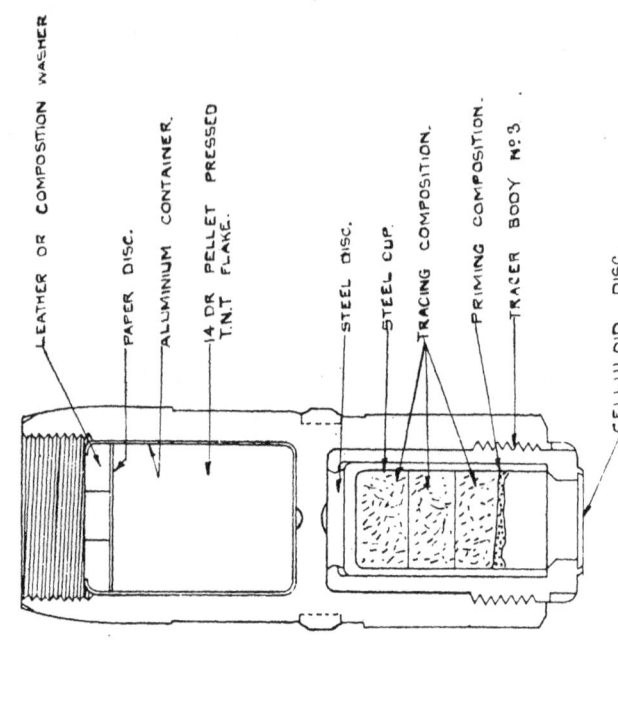

FIG. 20

shell from the normal H.E. shell. The brass case (or steel with a brass coating) is 9·8 inches in length and has the model number, "6331", and the designation of the equipment stamped in the base. The marking "Sprgr" is stencilled in white also on the base of the case. The length of the complete round is 13·4 inches and the weight, 2 lb. 13 oz. 8 drs.

The complete round consists of the following components :—
Fuze A.Z.39 (described in Pamphlet No. 4).
H.E./Tracer shell.
Tracer No. 3.
Case Model 6331 or 6331 St.
Propellant charge of double base composition with igniter.
Primer percussion C/13nA.

Shell (Sprenggranate 18 umg)

The shell with its upper and lower cavities separated by a diaphragm is the same as that described in Pamphlet No. 6. The aluminium container in the upper cavity contains a 14 dram pressed pellet of flake T.N.T. of density 1·56. The lower cavity contains a comparatively large tracer. The weight of the empty shell is 14 oz. 8 drs. The weight of the filled shell with fuze and tracer is 1 lb. 5 oz. 5 dr.

Tracer

A description of the No. 3 tracer is included in this pamphlet.

Case

The case is 9·84 inches long with an increase in taper at 9·25 inches from the base. The external diameters in front of the flange, at the shoulder and the mouth are 1·81, 1·57 and 1·52 inches respectively. The effective capacity is approximately 17·69 cubic inches (290 c.c.).

The designation of the equipment and the particulars of the propellant charge are stencilled in black on the side of the case. The marking "P.T.+25°C.", indicating the charge weight to be based on a standard charge temperature of 25° Centigrade, is also stencilled on the side of the case in red.

Propellant Charge

The weight, nature and size of the propellant charge, as marked on the case, is 164 grams of Digl. R.P.–8,2–(175 · 2,2/0,85). This marking is repeated on silk tape attached to the igniter bag. The propellant is in tubular form, the dimensions of the smooth black tubular sticks in inches being :—length 6·89, external diameter ·087, internal diameter ·033. The composition of the propellant, as found by analysis, consists of :—nitrocellulose 65·12 per cent. (Nitrogen content 12·14 per cent.), diethylene glycoldinitrate 31·52 per cent., centralite 2·65 per cent., potassium sulphate ·42 per cent. and ·29 per cent. of graphite.

The igniter consists of a circular silk bag secured to the base of the charge by a silk tape and containing 34·4 grains of the usual greyish green porous nitrocellulose powder in the form of short cylindrical grains. The mean dimensions of the grains are ·043 inch length and diameter. The composition, as found by analysis, consists of :—nitrocellulose (including graphite) 87 per cent., diethylene glycoldinitrate 11 per cent., centralite ·5 per cent., diphenylamine ·5 per cent., and 1 per cent. of volatile matter.

Primer

The primer percussion C/13nA is described in Pamphlet No. 7. In this instance, as in other of these primers recently examined, the amount of mercury fulminate in the cap composition has been decreased and granular gunpowder has been introduced between the cap and the pellet of pressed gunpowder. The cap composition consists of the following :—mercury fulminate 30·6 per cent., potassium chlorate 35·2 per cent., antimony sulphide 27 per cent. and 7·2 per cent. of ground glass.

GERMAN 3·7 cm. Pak HOLLOW CHARGE MUZZLE STICK BOMB
(3·7 cm. Stielgranate 41)
(Fig. 21)

The bomb is fired from the 3·7 cm. Pak (anti-tank gun) by inserting the rod at the tail of the bomb in the muzzle of the gun and using a separate Q.F. cartridge.

The streamlined body of the bomb has a dome shaped impact cap carrying a nose fuze and is fitted with a tail unit consisting of a steel rod surrounded by a perforated steel sleeve carrying three pairs of vanes. The overall length of the fuzed bomb is 29·1 inches and weight, filled and fuzed, is 19 lb. The exterior, except the rod, is painted deep olive green and stencilled in black. The stencilling includes the H.E. numeral " 95 " indicating the filling to be cyclonite/T.N.T. 60/40.

The complete round consists of the following components :—
    Fuze A.Z.5075.
    Bomb with tail unit.
    Two small gaines, Kl.Zdlg C/34 Np.
    Base fuze Bd. Z.5130.
    Cartridge case model 6331 St.
    Propellant charge of Ngl.R.P.−9,5−(188 × 2,5/0,9).
    Primer percussion C/13nA St.

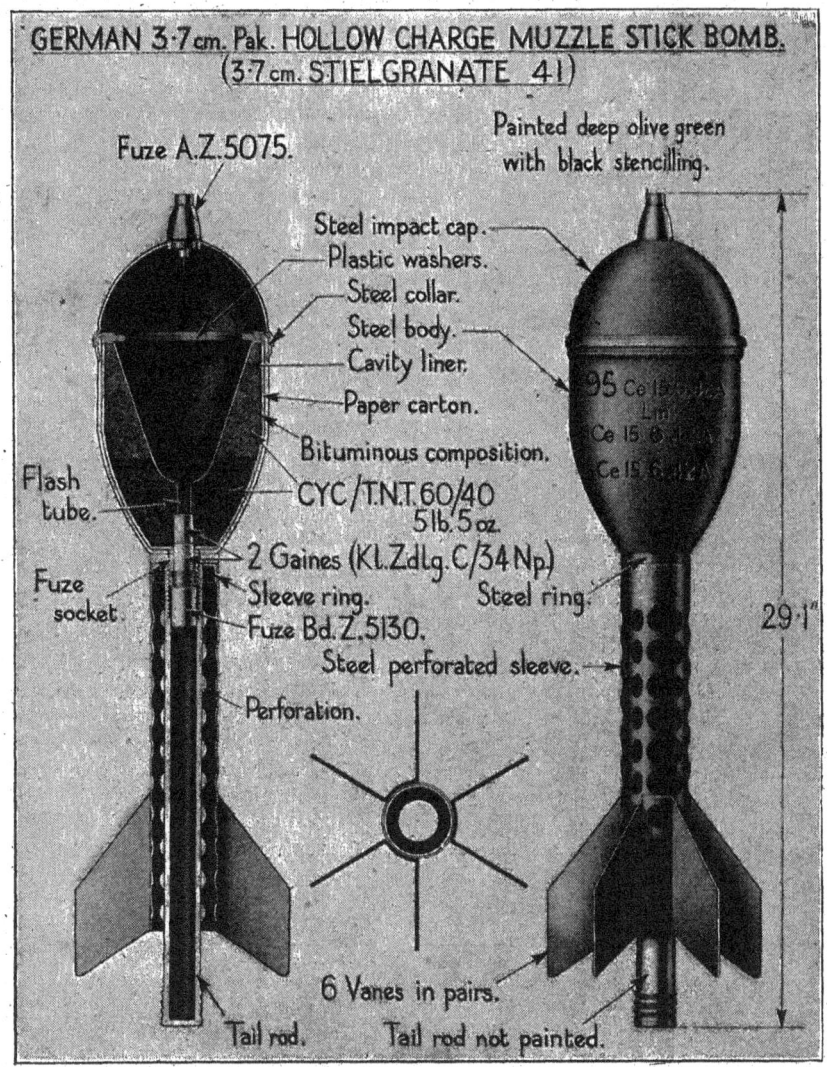

Fig. 21

Fuzes A.Z.5075 and Bd. Z.5130 and the gaine Kl.Zdlg. C/34 Np are described under separate headings in this Pamplet. Primer C/13nA St is described in Pamphlet No. 7.

The estimated performance against homogeneous armour is about 5 inches (127 mm.) at normal, and 3½/4 inches (89-102 mm.) at 30 degrees.

Bomb Body

The steel cup-shaped streamlined body has a thin wall and a flat base with a hole in the centre. Welded to the base is a steel ring carrying a fuze socket which fits into the hole in the base and protrudes below the ring. The socket is screwthreaded internally near the base for the insertion of the base fuze and is also screwthreaded externally for the assembly of the tail unit. The bursting charge, consisting of approximately 5 lb. 5 oz. of cyclonite/T.N.T. 60/40, is contained in a waxed carton and comprises two pellets. The pellets are shaped to fit closely in the body and to receive the cavity liner for the "hollow." The liner is of 2 mm. mild steel and is coned at approximately 40 degrees towards the base which is well rounded. A short metal flash tube, fixed in a hole at the base of the liner, leads to a cavity containing a gaine in the base of the lower pellet of the bursting charge. The liner has a small external flange at the top. The usual black bituminous composition secures the paper carton of the bursting charge to the body and the whole is retained by a thin steel collar which is rolled into the body near the top and has an internal rim overlapping a number of plastic washers assembled on the flange of the liner.

The steel dome shaped impact cap is cannelured near its base where it is supported by the collar to which it is attached by spot welding. A fuze socket is welded in a hole at the top of the cap for the insertion of the nose fuze.

The gaines are inserted through the socket for the base fuze. The upper gaine has its detonator towards the nose fuze and the lower, which is practically wholly contained in the socket, has its detonator towards the base fuze.

Tail Unit

The hollow steel tail rod is approximately 15 inches long and 1·42 inches in diameter. The base end is closed and near this end there are circumferential grooves which probably serve to obstruct the passage of the propellant gases over the rod. The upper end of the rod is screwthreaded internally for assembly to the fuze socket at the base of the bomb body. A steel ring welded at the top of the rod carries a steel perforated sleeve of approximately 2·4 inches diameter which surrounds the rod and extends down to three inches from its base. The perforations are of about ·75 inch diameter and extend throughout the length of the sleeve. Six vanes, formed in pairs, are attached near the base end of the sleeve.

Cartridge

The case is the normal 3·7 cm. Pak model 6331 St. closed at the mouth by a cork disc. The propellant charge consists of 216 grams of Ngl.R.P.–9,5–(188·2,5/0,9). This is a double base propellant of nitroglycerine and nitrocellulose in tubular form. The dimensions of the tubular cords are :—length 7·4 inches, external diameter ·098 inches, internal diameter ·035 inches.

The igniter at the base of the charge contains 10 grams of Nz.Man.N.P. (1,5·1,5). This is the nitrocellulose powder in the form of small cylindrical grains normally used in igniters.

GERMAN 8 cm. M.L. MORTAR H.E. BOMB 38
(8 cm. Wgr 38)
(Fig. 22)

The H.E. bomb, Model 38, for the 8 cm. mortar (Gr.W 34 8 cm.) contains an ejection charge inside its lightly attached head which, after impact, throws the body into the air where the bursting charge is detonated.

The bomb is painted deep olive green and is stencilled in black. The stencilling includes the H.E. numeral " 14 " on the head indicating a bursting charge of T.N.T. and the model number " 38 " below the weight class marking on the streamlined portion of the body. Externally the bomb has the same appearance as the dull red Model 34 bomb described in Pamphlet No. 6 but, in the absence of markings, it can be identified by the presence of four equally spaced pins in the second rib of the parallel portion and four similarly spaced grub screws in the head just above the parallel portion. A later model, Model 39, has only the four screws in the head. A bomb, filled and fuzed, bearing the weight class marking for Class II, weighed 7 lb. 5 oz. 8 drs.

Bomb

The bomb consists of three main parts, the head, the body and the tail unit. The tail unit is the same as that of the Model 34. The head extends down to the parallel portion of the bomb and fits over a rim formed on the closing plate in the top of the body. The head is secured to the side of the closing plate by four equidistant screws inserted from the exterior and contains an ejection charge in a celluloid capsule. The body, containing the bursting charge, is closed at the top by a circular closing plate which rests on an internal shoulder in the body and is secured by four pins also inserted from the exterior. The closing plate is recessed from the top having a circumferential rim which fits into the head. A screwthreaded hole is formed in the centre of the closing plate for the insertion of the delay chamber. The delay chamber consists of an inverted cup with a small central orifice in the closed upper end. Externally

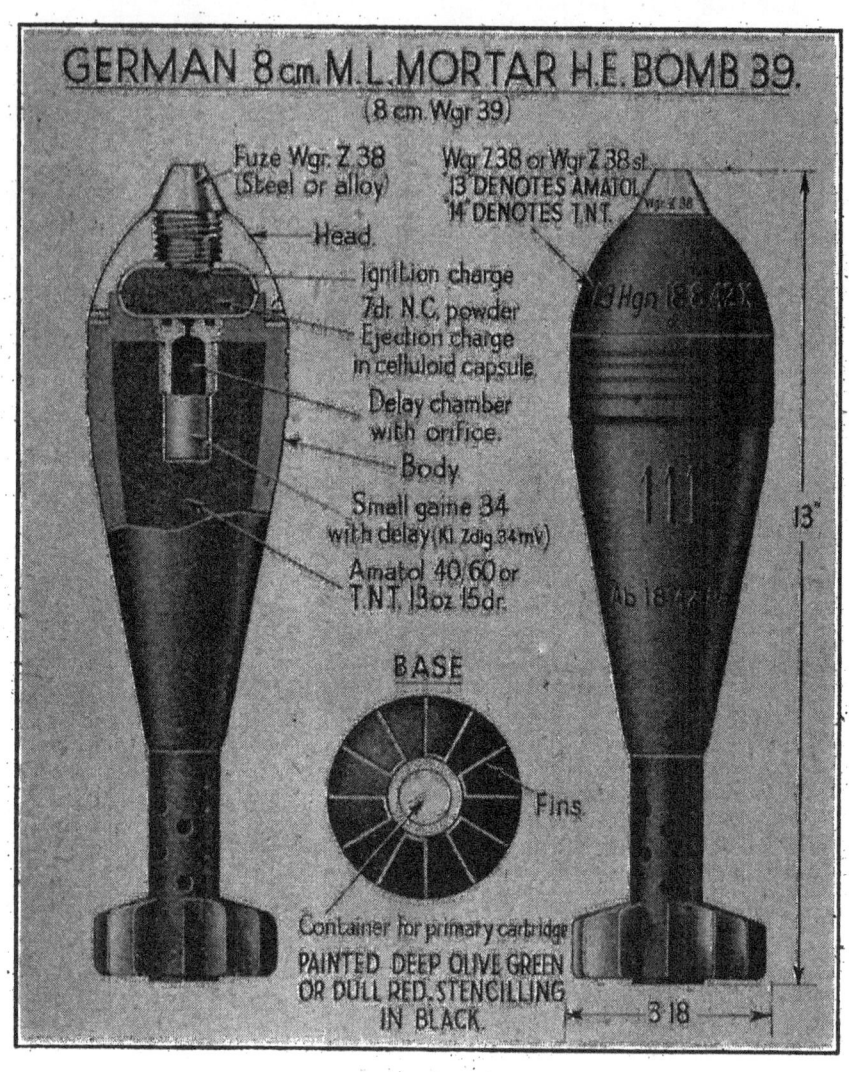

Fig. 22

the chamber is screwthreaded and flanged near the top for insertion in the closing plate and below this it is reduced in diameter and screwthreaded to receive an exploder container assembled at the base end.

Ejection Charge

The ejection charge in the head of the bomb, consisting of 12·32 grams of nitrocellulose powder in the form of porous green cylindrical grains, is contained in a celluloid capsule which is shaped to fit the cavity in the head and the recess in the closing plate. A celluloid cup set in the top of the capsule, beneath the fuze, contains an ignition charge consisting of pressed gunpowder, primed with a white ignition composition. The composition consists of lead thiocyanate, nitrocellulose, sulphur and potassium perchlorate and is also used to seal five holes in the base of the capsule above the delay chamber.

Bursting Charge

The bursting charge, contained in the body below the closing plate, consists of 13 oz. 12 drs. of T.N.T. with a short central cavity to receive the exploder container.

Gaine and Fuze

The small P.E.T.N. delay gaine (Kl.Zdlg. 34 m.V.) housed in the exploder container is described in this pamphlet as a separate item. The fuze, Wgr. Z.38, is included in the description of the 5 cm. bomb in Pamphlet No. 4.

Cartridges

The primary and augmenting cartridges were the same as those used for the 34 bomb (*see* Pamphlet No. 6) but four augmenting cartridges are now being used, instead of three, to increase the range of all 8 cm. bombs.

The composition of the 10 gram double base flake propellant (Ngl.Bl.P.-12,5-(1·1·02)) used in the primary cartridge consists of 59·92 per cent. of nitrocellulose (Nitrogen content 13·36 per cent.) and 39 per cent. of nitroglycerine and includes ·86 per cent. of ethyl centralite and ·22 per cent. of graphite. The initiator composition in the cap consists of 37·5 per cent. of lead styphnate, 4·2 per cent. of tetrazine, 7·4 per cent. of antimony sulphide, 12·4 per cent. of calcium silicide and 38·5 per cent. of barium nitrate. This composition is of approximately the same analysis as that of the " Sinoxid " cap known before the War.

The composition of the double base ring propellant in the augmenting or secondary cartridges consists of 60·61 per cent. of nitrocellulose (Nitrogen content 13·35 per cent.), 38·29 per cent. of nitroglycerine, ·7 per cent. of akardite and ·4 per cent. of graphite. The weight and designation of the propellant in each of the cartridges as indicated by the markings, is 9 grams of Ngl.Rg.P.12,5. The size, when three augmenting cartridges are provided, is (0,4-70/35). The size of propellant in the rounds provided with four cartridges is (0,4-60/30).

Action

On impact, the flash from the detonator of the fuze, with the assistance of the ignition charge in the top of the capsule, ignites the nitrocellulose ejection charge. The explosion of the charge ejects the body of the bomb from the head and, at the same time, initiates the gaine through the orifice in the top of the delay chamber. The delay resulting from the passage of the flash through the orifice and the chamber, also the delay arrangement in the top of the gaine, provides a time interval, reported to be about ·7 of a second, between impact and the detonation of the bursting charge. The height to which the bomb will be thrown in this time depends upon the nature of the ground. Bursts have been reported at heights between 20 and 60 feet (nature of ground not specified) but it is understood that the bomb is intended to detonate at heights between 3 and 20 feet.

GERMAN 8 cm. M.L. MORTAR H.E. BOMB 39
(8 cm. Wgr. 39)
(Fig. 23)

The H.E. bomb, Model 39, for the 8 cm. mortar (Gr.W.34 8 cm.), like the Model 38 bomb also described in this pamphlet, contains an ejection charge inside the lightly attached head which, after impact, throws the body into the air where the bursting charge is detonated.

The bomb is painted deep olive green or dull red and is stencilled in black. The deep olive green bombs encountered to date have been fitted with steel fuzes (Wgr.Z.38 St) and have been stencilled below the parallel portion of the body with the H.E. numeral " 14." This indicates a bursting charge of T.N.T. The dull red bombs have been fitted with the normal alloy fuze (Wgr.Z.38) and have the H.E. numeral " 13 " stencilled in the same position which indicates a bursting charge of Amatol 40/60. The bomb can be distinguished from the 34 and 38 models by the four small equi-distant screws securing the head to the body. These are found just above the grooved parallel portion of the bomb and are normally covered by a wax-like sealing composition.

By reference to the drawings it will be seen that the 39 Model differs from the 38, already described, in that instead of the body being closed at the top by a closing plate secured by pins, a diaphragm is formed integral with the body. Also, the celluloid capsule containing the ejection charge is of a different shape and is without the holes sealed with ignition composition in its base.

Cartridges

A primary cartridge and four augmenting cartridges of the types used with the earlier pattern bombs are used. The size of the

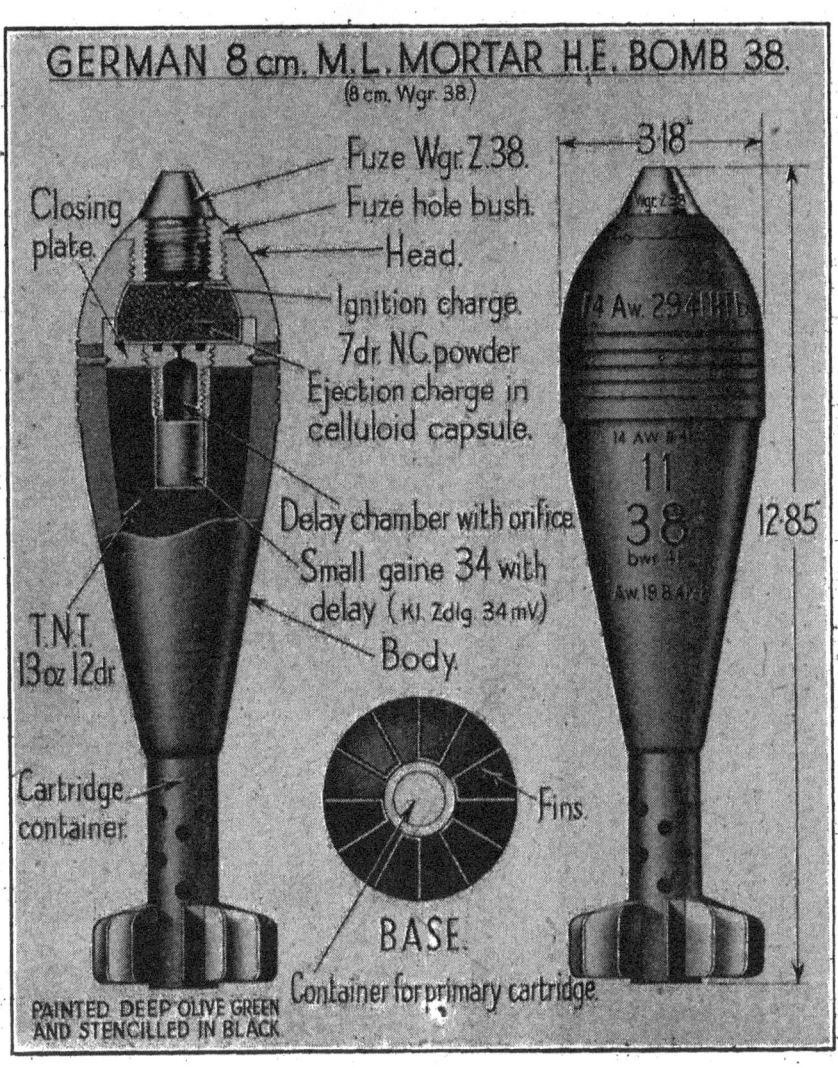

Fig. 23

propellant rings in the augmenting cartridges differs in that the external and internal diameters are 60 mm. and 30 mm. respectively, instead of 70/35.

GERMAN 10 cm. M.L. MORTAR (10·5 cm. CALIBRE) H.E. BOMB
(10 cm. Wgr 35)
(Fig. 24)

The bomb is fired from the 10 cm. Nb.W.35 (M.L. Smoke Mortar) with three charges consisting of a primary cartridge with one, two or four augmenting cartridges.

The streamlined bomb is fitted with a tail unit consisting of a perforated container for the primary cartridge and eight fins formed in pairs. The bomb is fuzed with the fuze Wgr.Z.38 and is painted deep olive green with black stencilling and a yellow band near the tail end of the body. The H.E. numeral " 14 ", stencilled on the head, indicates a bursting charge of T.N.T. The overall length, including the fuze is 17·1 inches and the total weight, filled and fuzed, is 16 lb. 11 oz. 12 drs.

The body of the bomb has a grooved cylindrical portion between the rounded head and the streamlined lower part and at this point is 10·44 cm. in diameter. A fuze-hole bush inserted in the nose carries an exploder container welded at its base.

Method of Filling

The bursting charge consists of 3 lb. 7 oz. 12 drs. of T.N.T. with a central cavity which contains a No. 8 smoke box under the exploder container which holds a large C/98 P.E.T.N. gaine. The smoke box, gaine and fuze are described in the following pamphlets :—

Smoke box (Rauchentwickler Nr8) Pamphlet No. 4, page 32.
Gaine (Gr. Zdlg C/98 Np) .. Pamphlet No. 6, page 14.
Fuze (Wgr Z.38) .. .. .. Pamphlet No. 4, page 7.

Cartridges

A cardboard distance cylinder ·55 inch in length is inserted in the cartridge container in the tail unit before the primary cartridge. The primary cartridge has the usual rolled paper body and is stamped " 10 cm " in the brass base. The cartridge is 2·88 inches in length and 1·04 inches in diameter at the mouth. The paper body is blue. According to the details given in a German handbook, the propellant charge consists of 4 grams of nitrocellulose powder in the form of small cylindrical grams (Nz.Man.N.P. (1,5-1,5) f.Wgr.) with the addition of 11 grams of nitrocellulose powder in the form of square flake (Nz.M.W.Bl.P.(2·2·0,45)).

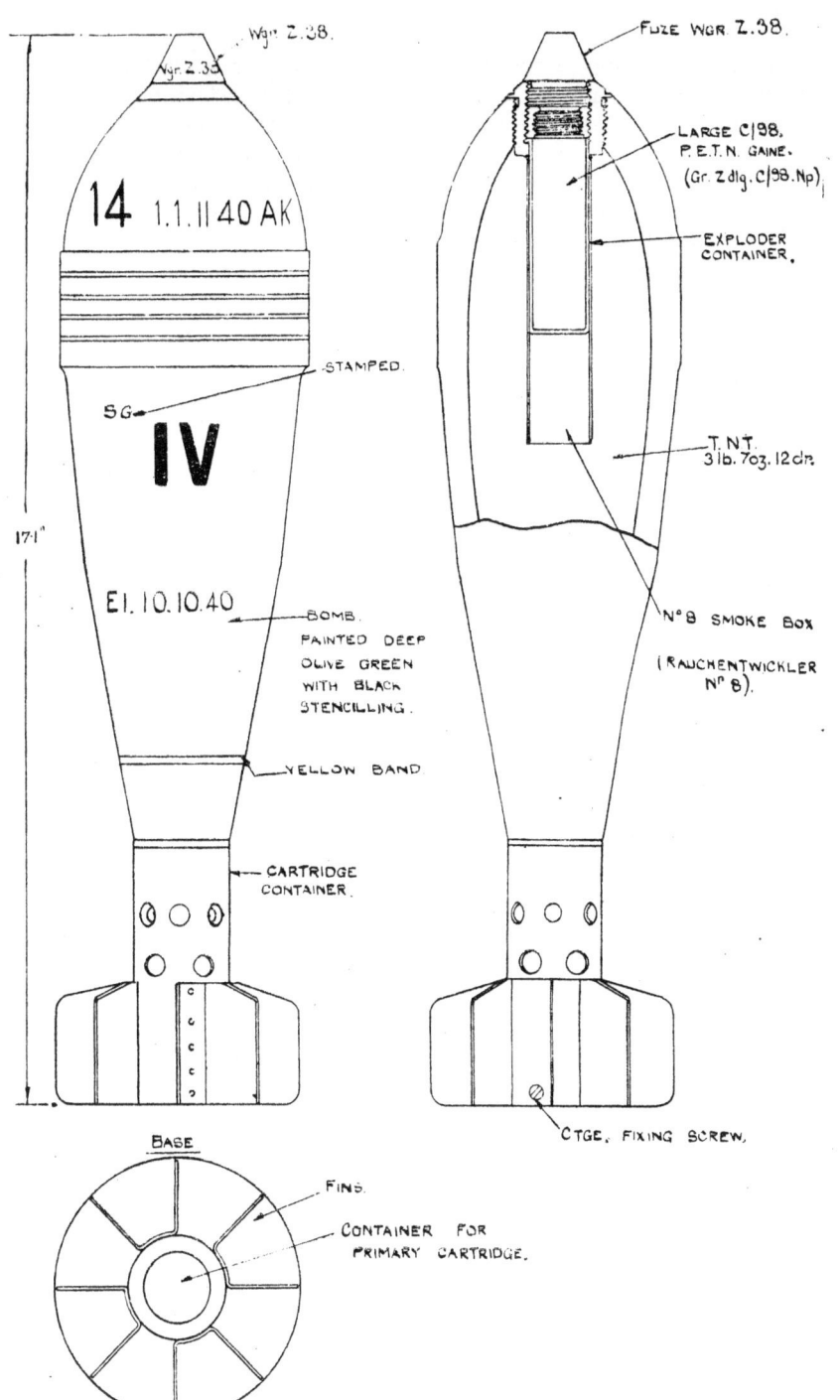

Fig. 24

The augmenting cartridges, four per bomb, are similar to those used in the 8 cm. mortar, each consisting of a flat ring-shaped bag containing rings of double base propellant (nitroglycerine and nitrocellulose). The bag and the rings have a radial split for the assembly of the cartridges above the fins on the tail unit of the bomb. The designation of the equipment, " 10 cm. Nb W " is stencilled on the bag. The weight, designation, shape and size of the propellant, as stencilled on the bag, are 21 grams Ngl. Rg.P.-12,5-(0,2-90/45).

GERMAN 10·5 cm. HIGH VELOCITY H.E. SHELL
(Fig. 25)

The shell consists of an 8·8 cm. streamlined body fitted with a hollow steel centering ring of 10·4 cm. diameter forming a shoulder and a corresponding base ring carrying the driving band both of which become detached during flight. The exterior is painted yellow and stencilled in black. The stencilling includes the H.E. numeral " 14 " near the nose, indicating a T.N.T. bursting charge, and the marking " R8 " at the approximate centre of the body. This marking indicates the inclusion of a No. 8 smoke box (Rauchentwickler Nr 8). With the optional delay fuze A.Z.23 (v.0·15), with which it is fuzed, the overall length is 17·5 inches and the total weight, 24 lb. 1 oz. 8 drs. Without the rings the weight is 20 lb. 10 oz. 8 drs.

Shell

The streamlined body of the shell has a comparatively thin wall and a head of large radius. Three equally spaced circular recesses are formed in the lower part of the head for the attachment of the centering ring. Just above the base of the shell, three corresponding circular recesses are formed to receive the inner ends of the steel cylinders keying the segments forming the base ring. This portion of the shell is grooved to engage a corresponding face inside the base ring and thus prevents longitudinal movement of the ring.

The steel centering ring consisting of a short cylinder with an internal flange at the base and a coned top is cut radially at three equi-distant points so that it is almost in three segments. On the interior of the cylindrical portion of the ring, midway between the cuts, three sockets are welded. The sockets are shaped at their inner ends to fit over a steel ball protruding from each of the recesses in the head of the shell and are screwthreaded internally to receive a grub screw. The band is assembled by placing it over the head of the shell and tightening the screws to engage the balls. The centering ring and fittings weighs 11 oz. 13 drs.

The base ring consists of three segments with a coned fairing, also in three parts, welded to its upper periphery. The segments are each keyed to the shell by a steel cylinder which passes through

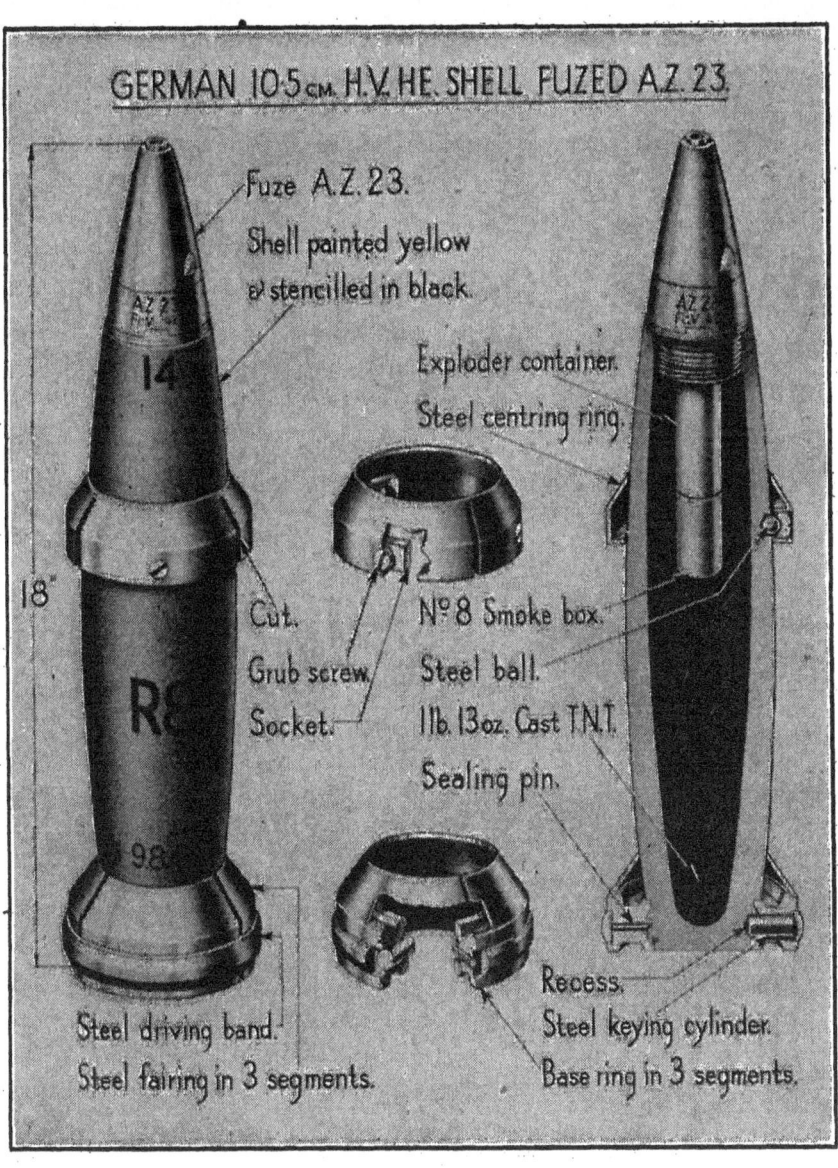

Fig. 25

a hole in the centre of the segment and is held between a recess near the base of the shell and the inner side of the driving band. The junction between the adjoining ends of the segments is sealed against the pressure of the propellant gases by a steel sealing pin which is inserted into a channel formed by coincident semi-circular grooves in the adjoining ends. The segments are retained in position by the steel driving band which is carried in a groove formed in the segments. The number " 783 " was stamped in the base of each of the segments on the shell examined. This stamping was probably used to guide assembly. The base ring and fittings weighs 2 lb. 11 oz. 3 drs.

A steel exploder container of the usual type is screwed into the fuze hole.

Method of Filling

The bursting charge consists of 1 lb. 13 oz. of cast T.N.T. with a central cavity for the exploder container and the No. 8 smoke box beneath the container. The gaine held in the container is the larger size of the C/98 model with a P.E.T.N. magazine filling. This gaine, the " Gr. Zdlg. C/98 Np ", is described in Pamphlet No. 6.

The smoke box, " Rauchentwickler Nr 8 ", is described under a separate heading in this pamphlet.

GERMAN 15 cm. HIGH VELOCITY H.E. SHELL

(Fig. 26)

This shell is similar to the 10·5 cm. H.V. shell described in this pamphlet and consists of a streamlined 12·8 cm. shell with a centering ring of 14·84 cm. diameter forming a shoulder and a corresponding base ring carrying the driving band. The construction of the base ring differs from that of the 10·5 cm. shell. Both of the rings are designed so that they become detached during the flight of the shell. The exterior is painted yellow and stencilled in black. The shell examined had no H.E. numeral stencilled on it but this was stamped, in accordance with the usual practice, near the fuze hole. The normal marking for the T.N.T. bursting charge would be the numeral " 14 " stencilled in black near the fuze hole. The marking " R 9 ", indicating the inclusion of a No. 9 smoke box (Rauchentwickler Nr 9) is stencilled prominently on the head above the centering ring. The shell examined also bore the stencilling " Z 3033 " as shown in the drawing. The meaning of this marking is not apparent.

With the optional delay fuze A.Z.23 (v. 0·15), with which it is fuzed, the overall length is 25·6 inches and the total weight 65 lb. Without the rings the weight is 57 lb. 15 oz. 12 drs.

Shell

The streamlined body has a comparatively thin wall and a head of large radius. The method of attachment and the construction of

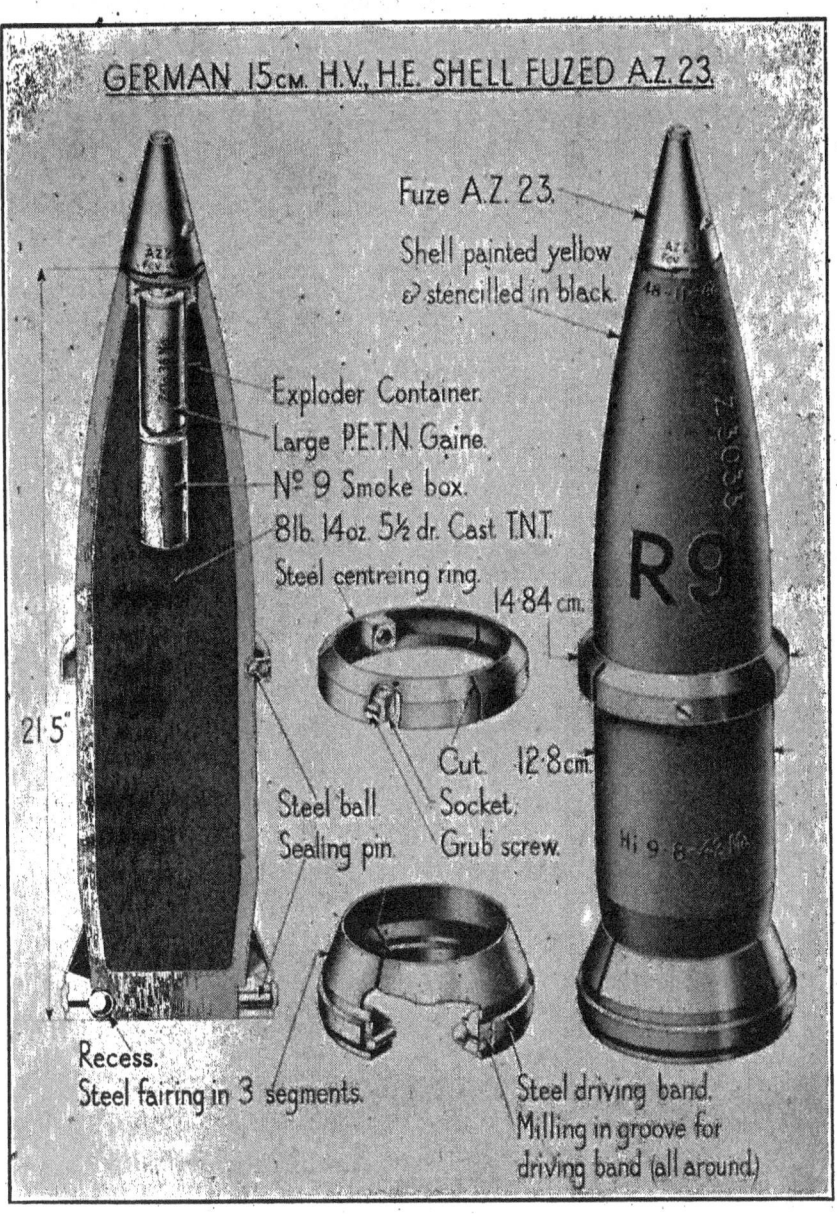

Fig. 26

the centering ring is the same as that described for the 10·5 cm. shell. At the base the body is reduced in diameter for the assembly of the base ring and has three equally spaced recesses each of which contains a protruding steel ball for the keying of the base ring. The centering ring and fittings weigh 14 oz. 11 drs.

The base ring, consisting of three segments with a coned fairing in three corresponding parts welded to it, has an interior shoulder which engages a shoulder near the base of the shell to prevent longitudinal movement. Each of the segments has a recess at the centre of its inner side to engage the steel balls protruding from similar recesses in the shell. A small radial channel leads from the recess to the exterior of the segment. The junction between ends of adjoining segments is sealed against the pressure of the propellant gases by a steel pin inserted into a channel formed by coincident semi-circular grooves in the adjoining ends. The segments are held in position around the base of the shell by the steel driving band. The groove for the driving band, formed in the segments, has a band of milling protruding from the centre to prevent independent rotation of the band. Each of the segments on the shell examined had the number " 156 " stamped in its base. This number was also stamped in the body about two inches above the fairing. The three portions of the fairing were stamped " 273 ", " 274 " and " 275 " respectively. These numbers were also stamped in the shell above the respective portions of the fairing and are probably a guide for assembly. The base ring and fittings weigh 6 lb. 1 oz. 7 drs.

A steel exploder container, larger in size than that in the 10·5 cm. shell, is screwed into the fuze hole.

Method of Filling

The bursting charge consists of 8 lb. 14 oz. 5½ drs. of T.N.T. cast with a central cavity to accommodate a No. 9 smoke box under the exploder container. The container holds a large P.E.T.N. gaine, the " Zdlg. 36 Np ", which has a convex base and has a filling about three times the weight of that in the Gr. Zdlg. C/98 Np. The weight of the filling is 107·5 grams. The large C/98 Np gaine contains approximately 35 grams.

The smoke box, Rauchentwickler Nr 9, is described under a separate heading in this pamphlet.

www.ingramcontent.com/pod-product-compliance
Lightning Source LLC
Chambersburg PA
CBHW032012080426
42735CB00007B/581